URBAN GOVERNMENT FINANCE

Volume 20, URBAN AFFAIRS ANNUAL REVIEWS

INTERNATIONAL EDITORIAL ADVISORY BOARD

ROBERT R. ALFORD
University of California, Santa Cruz

HOWARD S. BECKER
Northwestern University

BRIAN J. L. BERRY
Harvard University

ASA BRIGGS
Worcester College, Oxford University

JOHN W. DYCKMAN
University of Southern California

T. J. DENIS FAIR
University of Witwatersrand

SPERIDIAO FAISSOL
Brazilian Institute of Geography

JEAN GOTTMANN
Oxford University

SCOTT GREER
University of Wisconsin, Milwaukee

BERTRAM M. GROSS
Hunter College, City University of New York

PETER HALL
University of Reading, England

ROBERT J. HAVIGHURST
University of Chicago

EHCHI ISOMURA
Tokyo University

ELISABETH LICHTENBERGER
University of Vienna

M. I. LOGAN
Monash University

WILLIAM C. LORING
Center for Disease Control, Atlanta

AKIN L. MABOGUNJE
University of Ibadan

MARTIN MEYERSON
University of Pennsylvania

EDUARDO NEIRA-ALVA
CEPAL, Mexico City

ELINOR OSTROM
Indiana University

HARVEY S. PERLOFF
University of California, Los Angeles

P.J.O. SELF
London School of Economics and Political Science

WILBUR R. THOMPSON
*Wayne State University and
Northwestern University*

URBAN GOVERNMENT FINANCE
Emerging Trends

Edited by
ROY BAHL

Volume 20, URBAN AFFAIRS ANNUAL REVIEWS

 SAGE PUBLICATIONS   Beverly Hills   London

Copyright ©1981 by Sage Publications, Inc.

All rights reserved. No part of this book may be reproduced or utilized in any form or by any means, electronic or mechanical, including photocopying, recording, or by any information storage and retrieval system, without permission in writing from the publisher.

For information address:

    SAGE Publications, Inc.   SAGE Publications Ltd
    275 South Beverly Drive   28 Banner Street
    Beverly Hills, California 90212   London EC1Y 8QE, England

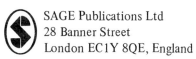

Printed in the United States of America

Library of Congress Cataloging in Publication Data

Main entry under title:

Urban government finance.

  (Urban affairs annual reviews ; v. 20)
  Includes index.
  1. Finance, Public—Addresses, essays, lectures.
2. Local finance—Addresses, essays, lectures.
3. Municipal finance—Addresses, essays, lectures.
I. Bahl, Roy W.   II. Series.
HJ236.U6   352.1   80-39559
ISBN 0-8039-1564-0
ISBN 0-8039-1565-9 (pbk.)

FIRST PRINTING

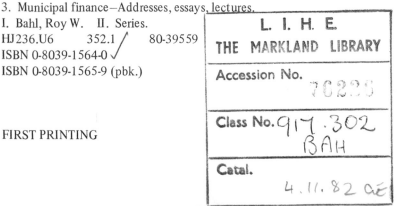

CONTENTS:

Introduction □ *Roy Bahl*   7

1 □ The Economy and the Federal Budget in the 1980s: Implications for the State and Local Sector □
Robert Reischauer   13

2 □ Public Employment's Impact on the Future of Urban Economies □
Shawna Grosskopf   39

3 □ The Prospects for Urban Revival □
James W. Fossett and Richard P. Nathan   63

4 □ The Fiscal Outlook for Growing Cities □
Bernard L. Weinstein and Robert J. Clark   105

5 □ The Tax and Expenditure Limitation Movement □
Deborah Matz   127

6 □ The Location of Firms: The Role of Taxes and Fiscal Incentives □
Michael Wasylenko   155

7 □ The Next Decade in State and Local Government Finance: A Period of Adjustment □
Roy Bahl   191

8 ☐ Fiscal Problems and Issues in Scandinavian Cities ☐
Joergen R. Lotz   221

9 ☐ Urban Finances in Developing Countries ☐
Johannes F. Linn   245

The Contributors   285

Introduction

☐ THE OUTLOOK FOR STATE AND LOCAL GOVERNMENT finance at the end of the 1960s must have been reasonably optimistic. The economy had grown steadily throughout the decade, the inflation rate remained low, general revenue sharing had been institutionalized as an important local revenue source, and the most serious of city fiscal problems were softened by state and local government revenue growth. The major problem was what to do about the decaying inner cities in the older region and the centerpieces of an implicit national urban policy were new towns and urban renewal.

In fact, the 1970s brought major crises in state and local government finance and a substantial change in the structure of federal-state-local relations. The decade was characterized by two recessions—the second far more serious than the first—and a high rate of inflation. The recessions and recoveries had uneven effects on the people and governments in the urban areas of the North and South and the movement of population and economic activity to the sunbelt continued. For the first time since the Great Depression, cities defaulted and other state and local governments teetered on the brink.

In response to the fiscal crises of the 1970s, the fiscal policies of state and local governments, as well as federal intergovernmental policy, changed dramatically from the pattern in the previous decade. Public employee layoffs were one response; public employee compensation deferrals, the postponing of capital projects, and the elimination of government programs were others. By the end of the 1970s, retrenchment had become the watchword of state and local government budgeting in the declining region. The growing region, healthier and less affected by national economic changes—but perhaps wary of the fiscal implications of continued growth—moved to adopt fiscal limitations.

Federal policy toward the state and local government sector during the decade was ambiguous, often contradictory, and generally ill-conceived. Total grants-in-aid quadrupled, but the net fiscal relief and economic stimulus effects of the hundreds of grant programs could not be evaluated. A "stimulus package" of federal grants shifted the focus to federal-local relations with the large public employment (CETA), counter-cyclical (ARFA), and public works programs. But by the end of the decade, the federal government was at least thinking about balanced budgets, and the rapid growth in grants-in-aid was over, at least temporarily.

What of the outlook for state and local government finances in the 1980s? The essays in this volume assess the prospects and speculate at the possible policy responses in a context of the recent history outlined above.

Robert Reischauer considers the national economic and federal budget prospects for the 1980s and their implications for the state and local government sector. His prognosis is gloomy—most forecasts suggest that the 1980s will be characterized by relatively slow economic growth, extremely high rates of inflation, high levels of unemployment, rapid nominal wage growth, and high interest rates. This poor performance of the economy plus the possible success of efforts to curb federal spending and shift budget priorities away from social programs could have profound effects on the fiscal health of state and local governments. Shawna Grosskopf discusses one aspect of this effect—that on the growth of public employment and public employee compensation. She finds the existing theoretical models not particularly helpful in explaining the behavior of public employment during the 1970s or for predicting behavior in the 1980s, because they were constructed to explain sectoral behavior in periods of growth rather than decline. From historical trends, however, she has sorted out some aspects of the behavior of state and local government employment over the business cycle and during the slow growth period of the 1970s: the state and local sector was mildly employment stabilizing, and compensation rates have fallen behind those in the private sector during the last half of the 1970s. She reads these trends to suggest that there will be a continued shift away from public employment within the broader "service" sector.

The next two chapters consider a major policy issue for the 1980s: defining and measuring fiscal distress. James Fossett and Richard Nathan index urban hardship by concentrating on city age, population loss, and poverty. Their resulting "urban condition" indices show that the older cities of the Northeast and industrial midwest are generally worse off, and that their *relative* position has actually worsened. An alternate construct, an economic performance index based on value added, gross sales, and employment structure, reaches a similar conclusion. The implications of their findings are that with slow-growing economies, the fiscal outlook for such cities is grim. Bernard Weinstein and Robert Clark take a different view in an essay on the fiscal outlook for growing cities. They develop the case that declining populations and economic bases do not necessarily result in fiscal distress, and that cities experiencing demographic and economic improvements may well be in financial troubles precisely because of this growth. They speculate that the growing metropolitan areas of the South and West may realize slow or negative population growth and increased fiscal pressures during the 1980s. The controversy over what causes fiscal distress and over how federal monies ought to be targeted, promises to be an important and lively debate in the coming years.

Deborah Matz and Michael Wasylenko consider two of the major state and local government tax policy responses to the fiscal problems of the 80s. Matz reviews the fiscal limitation movements of the late 1970s and offers a useful analysis of how they are likely to affect aggregate state and local government spending and service levels in the coming decade. She points out that the limitation movement has been concentrated in the growing region, that limits are usually tied to income growth, and that, therefore, a drastic reduction in the overall level of state and local government spending is unlikely. Still, she expects the effects in the 1980s to be a minimizing of the implementation of new programs and an increase of the use of user charges. Wasylenko takes on the equally important issue of the role of taxes and fiscal incentives in firms location. He concludes that taxes play relatively little role in interregional migration but may play a more important role in intraregional business location decisions. More importantly, he points out that states should not ignore business location considerations in making tax policy, because the lack of any fiscal inducements may cause firms to suspect a bad business climate. In all likelihood, the 1980s will see further competition among the states in using fiscal incentives to attract industry.

I attempt to put together the list of factors affecting the outlook for state and local government finance and the possible national urban policy response, in order to discuss the difficult adjustments that surely lie ahead. The 1980s will be a time of substantial adjustment as the older cities of the North continue to retrench, reduce *relative* public service levels, and, at least for the time being, reduce their dependence on federal grant revenues. On the other hand, the growing regions will be involved in a "catch up" when they have to face backlogs of public services, the demands for better services from new residents, and the demands of state and local government employees whose wages are low. All of this will be compounded by inflation. National urban policy ought to define the federal response to these problems, in particular with respect to four issues: whether a revitalization or a compensation strategy will be followed in assisting cities, whether the business cycle will be explicitly recognized in federal intergovernmental policy, how the federal government will intervene in the face of big-city financial emergencies, and what future role will be given to state government.

The two final chapters offer a comparative dimension by considering urban fiscal structure and problems elsewhere in the world. Joergen Lotz describes some striking similarities between the problems of American cities and those of the large Scandanavian cities. Interestingly, however, he describes a response to central city decline that has long been advocated by urban scholars in the United States—the taxation of local income and the creation of a needs-based grant system. Johannes Linn provides an excellent description of the relatively unknown area of urban government finances in developing countries. Though the Third World cities have their substantial population size in common with American urban areas, their problems are quite different. The principle issue seems to be the almost impossible backlog of public services to be filled and the very great restrictions on local fiscal autonomy. An interesting lesson from the developing country experience, however, is that where this central governmental control has been relaxed, innovative and imaginative local governmental fiscal policies have sometimes emerged.

These essays only scratch the surface of a long list of important issues to be dealt with in formulating policy for the state and local government sector in the 1980s. It is important to realize that many of the issues to be considered will be new—at least for a few years. The state and local government sector will be much less the growth industry than it has been in the past.

—*Roy Bahl*

1

The Economy and the Federal Budget in the 1980s: Implications for the State and Local Sector

ROBERT REISCHAUER
Congressional Budget Office

☐ DURING THE PAST DECADE, the state and local sector was greatly affected by the national economy and by changes in governmental policy as expressed in the federal budget. This chapter briefly reviews past economic and budget trends and examines likely future developments to isolate those factors that could affect significantly the state and local sector during the 1980s.

THE ECONOMY

THE PAST DECADE

The economic performance of the United States deteriorated markedly during the 1970s. High inflation, high unemployment, lagging productivity, and an unfavorable international balance of merchandise trade characterized the period. Compared to the 1960s, the decade of the 70s saw the inflation rate almost triple, the unemployment rate rise by over a quarter (29%), the rate of productivity growth decline by half, and the balance of merchandise trade shift from a $4.1 billion surplus to a $10.4 billion deficit (see Figure 1.1).

AUTHOR'S NOTE: *The views expressed in this chapter are those of the author and do not represent positions of the Congressional Budget Office.*

14 URBAN GOVERNMENT FINANCE

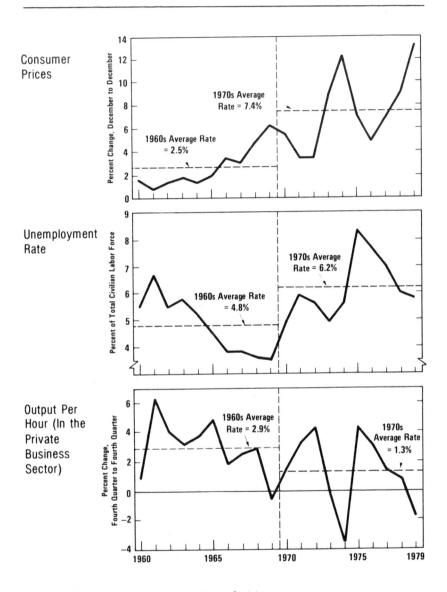

SOURCE: U.S. Department of Labor, Bureau of Labor Statistics.

Figure 1.1 Measures of Economic Performance, 1960-1979

The explanations for this pronounced deterioration are fairly straightforward. Inflation was driven up by excess demand pressures, external shocks, governmental policies, and the growing ability of groups to protect their real wages despite lagging productivity growth. Excess demand pressures arose in the late 1960s and early 1970s largely because federal policymakers were unwilling to present the American people with the bill for both the Vietnam War and the domestic initiatives of the Great Society and New Federalism eras. While these excess demand pressures signaled the end to the period of under 2% annual inflation—an era that ran from 1958 through 1965—they played little part in pushing the inflation rate up to double-digit levels.

The original impetus for double-digit inflation was a combination of external shocks. Bad weather, poor harvests, and the disappearance of the anchovies off the coast of Peru combined to push agricultural prices up by over 50% between 1972 and 1974. The Arab oil embargo of 1973 was followed by wholesale energy price increases of 60% in one year. Industrial input prices rose by 69%, spurred on by shortages of certain raw materials and capacity bottlenecks. The result was the double-digit inflation rate of 1974 (11.0%), the first double-digit rate experienced by the United States since 1947. Other shocks have followed, the latest being the turmoil in Iran that facilitated an increase in the price of imported oil of 120% during 1979. The indirect and direct effects of this increase alone added 2.6 percentage points to the Consumer Price Index (CPI) in 1979.[1]

Governmental policies exacerbating inflation during the decade included higher agricultural price supports, environmental legislation, automobile safety and emission standards, deregulation of natural gas and domestic oil prices, laws regulating worker safety (OSHA), pension reform laws (ERISA), trade restrictions on such items as imported TVs and shoes, and legislation to make public places and institutions more accessible to the handicapped (Crandall, 1978). These policies had worthwhile objectives, but few considered their inflationary costs.

Additional governmental actions pushing up the cost of living include higher Social Security taxes and increased minimum wages, both of which are believed to be partially passed on to consumers in the form of increased product prices. The Social Security taxes paid by employers have risen from 4.8% of the first $7,800 of wages per worker in 1970 to 6.13% of the first $25,900 of wages in 1980. The minimum wage has risen from $1.45 per hour in 1970 to $3.10 per hour in 1980. Social Security tax increases were needed to fund the new Medicare program, the explosion in the Disability Insurance rolls, and the increase in elderly OASI recipients.

Monetary policies also fed inflation in the late 1970s. In addition to the rapid increase in the money supply over the last half of the decade, the Federal Reserve's credit-tightening actions of November 1978, October 1979, and February and March 1980 forced interest rates up significantly. Increased mortgage interest rates alone pushed up the CPI by 1.6 percentage points in 1979. The general weakening of the dollar during the decade contributed to inflation by increasing the prices of imported products.

In part, inflation persisted during the 1970s despite periods of weak demand, because significant segments of society—workers and nonworkers—have been able to protect their incomes from erosion by past inflation. Social Security checks, food stamps, and veterans' benefits, among others, are automatically increased—that is, indexed—when the cost of living rises. Workers in the highly unionized sectors of the economy—automobiles, primary metals, machinery, rubber, chemicals, paper, petroleum, transportation, and utilities—have been protected by cost-of-living escalators. Often their wages have grown with little apparent relationship to productivity increases or the strength of the market for the product produced.

An example of this is the steel industry, where wages rose by about 10.5% annually through the decade, well above the increase for the manufacturing sector as a whole. This occurred despite the high initial level of steelworkers' wages, low industry profits, significant inroads made by imported steel, the closing of some steel plants, and job losses. Had steelworkers' wages risen during the decade at the average rate for all workers in the manufacturing sector, the input costs of steel would be roughly

11% lower today. Such higher labor costs have played a major role in sustaining high inflation rates.

High unemployment during the 1970s resulted from demographic changes in the labor force and the government's reluctance to pursue highly expansionary fiscal policies. The labor force grew rapidly—by 24.4% in the 1970s; the rate was 16% in the 1960s. Much of this growth consisted of women (59%) and younger workers, groups traditionally experiencing high rates of unemployment. These demographic shifts have led many economists to raise their estimates of the "full employment" or noninflationary unemployment rate—that is, the rate below which inflation is thought to accelerate—from the 4.0% to 4.5% range to the 5% to 6% range (Wachter, 1976a, 1976b). The government's reluctance to bring the unemployment rate down through highly expansionary fiscal policies has stemmed largely from the fear of exacerbating the already-high rate of inflation. Some have questioned the feasibility of lowering the unemployment rate significantly, because whenever jobs are readily available, the number of new entrants drawn into the force has tended to surge.

The decline in productivity growth during the decade is not fully understood (Denison, 1979). The changing composition of the work force, sectoral shifts in the economy, and the depressed level of business fixed investment are some of the factors thought to be responsible for this slowdown. The influx of new, younger and secondary workers—groups with little work experience and low productivity—into the labor force probably dampened productivity growth. The continued shift away from the high productivity agricultural and manufacturing sectors toward services having low measured productivity has had the same effect. The slowdown in the growth of capital per worker has also lowered productivity growth. Real business fixed investment—from which much of the increase in productivity must come—grew at 2 1/2%, or 4 1/2 percentage points slower than the previous decade. This was despite the fact that significant amounts of investment were required during the decade for pollution control equipment and other "nonproductive" purposes. The rapid increase in energy costs, the spread of governmentally imposed environmental and safety regulations, a slower rate of technological innovation, and reduced real spending on research and development are suggested as additional possible causes for slower growth in output per worker. The slow-down in productivity growth weakened the chief means of breaking the wage-price spiral. The final characteristics of the economy during the decade were the deteriorations in the U.S. merchandise trade position and in the value of the dollar. These were attributable largely to oil imports. Agricultural exports were generally strong during the decade, and the relative decline of manufactured exports slowed considerably in the 1970s from the 1960s.

THE DECADE OF THE 1980s

Whether the economic deterioration of the 1970s is a temporary aberration or is just the precursor of an even worse decade is a question impossible to answer except in the most general and uncertain way. Certainly the decade of the 1980s is off to an inauspicious start. Inflation raged ahead at an 18% annual rate during the first quarter of 1980, and increased at a double-digit rate for the year as a whole. The unemployment rate is projected to remain well over 7%, as the economy experiences a period of prolonged weakness. Of course, unforeseen domestic political and international events will determine the course of the 1980s just as they did the 1970s. Nevertheless, one can paint a general picture of the economic potential.

While they differ on the specifics, most long-run forecasts suggest that, by historical standards, the 1980s will be characterized by relatively slow economic growth, extremely high rates of inflation, high levels of unemployment, rapid nominal wage growth, and high interest rates (Table 1.1). The figures in parentheses are projections assuming that the economy follows a path characterized by the large cycles of the past decade; the other figures are projections assuming that the cycles are small.

The sluggish rate of economic growth will result largely from slowdown in the growth of the labor force. The growth in the labor force will slacken because most of the postwar babies will have entered the labor market and because the labor force participation rate of married women is not expected to increase as rapidly in the 1980s as it did in the 1970s. Energy shortages and the possibility of significant investment channeled into economically inefficient investment, such as synthetic fuels production, could further slow growth.[2]

Inflation, measured either by the CPI or by the GNP deflator, should be higher than at any time in recent history. This gloomy projection rests not on the prediction of new external shocks, but rather on the rates of price increase woven into the fabric of the economy and into the expectations of workers and consumers during the past few years and on the inability of institutions and policies to wring this inflation out of the economy.

The growth of productivity should recover from the dismal pace of the last half decade, in part because the labor force will age, increase its skills, and become more productive, and in part because the stock of plant and equipment per worker should rise. Nevertheless, it is not likely to attain the rates experienced during the 1950s and 1960s. Despite low productivity growth, however, nominal wage should rise rapidly. Real compensation gains and real disposable income growth should be slow and possibly negative in some years, as inflation and taxes eat into the purchasing power of paychecks. Unemployment should remain high for the same reasons it has been high in recent years.

THE FEDERAL BUDGET

THE PAST DECADE

During the 1970s, the federal budget was characterized by a rapid growth in spending, persistent deficits, significant shifts in priorities, and increasing uncontrollability.

Federal outlays grew rapidly in nominal terms, in real terms, and relative to the size of the economy. Nominal spending grew 187% in the 1970s; growth during the 1960s was 113%. Most of this growth could be attributed to inflation. Nevertheless, federal spending rose some 30% after adjusting for inflation. This real growth was significantly below the real growth rate of 46% experienced during the previous decade, nevertheless it was faster than the growth of the economy as a whole. As a result,

TABLE 1.1 DRI Long-Term Projections (average annual percentage change)

Economic Variables	1956–1960	1961–1965	1966–1970	1971–1975	1976–1980	1981–1985	1986–1990
Economic Growth: Real GNP	2.4	4.7	3.0	2.3	3.3	3.4 (3.4)	2.9 (1.7)
Prices							
CPI	2.0	1.3	4.3	6.8	9.0	9.2 (10.5)	8.2 (8.5)
GNP Deflator	2.4	1.6	4.2	6.8	7.3	8.7 (9.3)	8.1 (8.7)
State and Local Purchases Deflator	3.5	2.8	6.3	8.0	8.4	9.0 (9.6)	8.8 (9.5)
Unemployment Rate (percentage)	5.3	5.5	3.9	6.1	6.8	7.3 (7.5)	5.8 (6.4)
Real Disposable Income	2.7	4.7	3.9	3.0	2.8	3.0 (2.9)	2.9 (2.5)
Productivity: Output per Hour (Nonfarm Business Sector)	1.7	3.5	1.4	1.5	0.5	1.3 (1.1)	1.7 (1.0)
Wages: Compensation Per Man Hour	4.8	3.8	6.4	8.0	8.6	10.2 (10.7)	9.6 (10.1)
Prime Interest Rate (percentage)	4.2	4.5	6.7	7.5	9.8	11.0 (11.9)	10.8 (11.0)
State and Local AAA Bond Rate (percentage)	3.0	3.1	4.6	5.5	5.8	6.1 (6.4)	5.7 (6.0)

SOURCE: Data Resources (1980).
NOTE: Figures in parentheses are based on the assumption that the economy follows cyclical patterns similar to those of the last decade; the other figures are based on the assumptions that cyclical swings will be smaller.

federal spending, as a percentage of the GNP, rose. While federal spending constituted 19% of GNP in the early 1960s, it amounted to 22% during the last half of the 1970s. This can be seen in Table 1.2, which shows five-year averages of the outlays, receipts, and deficits of the federal government as a percentage of the GNP. Five-year averages are provided to reduce the cyclical distortions evident in annual data. Such distortions occur because federal spending increases and federal revenues decrease automatically when the economy (GNP) weakens.

Since 1960, the federal budget has experienced a deficit in every fiscal year but one (1969). The average size of these deficits has grown in nominal and real terms and relative to the economy. Since 1975, the deficit has averaged $47.6 billion, and roughly one-third of the present national debt has been accumulated since 1975. The persistence and growth of these deficits can be explained by weak economic growth and by the fact that the Congress and the administration have been unwilling to let revenues (taxes) increase significantly as a fraction of national income.

In priorities, there has been a dramatic shift in federal spending away from defense and toward human resources. This shift reflects a continuing trend that began in the late 1950s (see Table 1.3). In 1955, defense spending absorbed 58% of the budget, while just over one-fifth of the budget went to support human resources programs. In 1980, less than one quarter of federal outlays were devoted to national defense, while over half went for health, education, income security, employment and training, and other human resource programs. In large measure, this shift in spending priorities reflects the creation of new programs; food stamps, Medicare, Medicaid, supplemental security income (SSI), Basic Educational Opportunity Grants (BEOGs), and Section 8 low-income housing were all initiated after the mid-1960s.

Grants to state and local governments, a category split between the "Human Resources" and "All Other" groups in Table 1.3, has also grown rapidly in importance. They now run over $88 billion a year, up from $24 billion in 1970 and $7 billion in 1960. Their relative importance to states and localities has also increased—from 19.4% of their spending in 1970 to 25.3% in 1980 (U.S., OMB, 1980). This growth is largely the result of new programs. The 1960s and 1970s saw the passage of General Revenue Sharing, Urban Development Action Grants (UDAG), the Community Development Block Grants (CDBG), Law Enforcement Assistance (LEAA) grants, urban mass transit grants, the Comprehensive Employment and Training Act (CETA), and Medicaid, to name but a few of the larger programs. Not only did the federal government assist states and localities in performing nonfederal functions, but it also used these governments as contractors to

TABLE 1.2 Federal Receipts, Outlays, and Deficit as a Percentage of GNP, FISCAL YEARS 1955-1979 (Five-Year Averages, Unified Budget)

	Receipts	Outlays	Deficit
1955-1959	17.2	17.9	0.7
1960-1964	18.4	19.1	0.7
1965-1969	18.9	19.8	0.8
1970-1974	19.2	20.4	1.2
1975-1979	19.4	22.0	2.6

SOURCE: U.S. Office of Management and Budget (1980a).

TABLE 1.3 Distribution of Federal Outlays by Broad Category, Fiscal Years 1955-1980 (Percentage of All Outlays)

Category	1955	1960	1965	1970	1975	1980
National Defense	58.1	49.0	40.1	40.0	26.6	23.1
Human Resources	21.2	27.7	29.9	37.4	51.7	53.0
Net Interest	7.1	7.5	7.2	7.3	7.1	9.2
All Other	13.5	15.8	22.8	15.4	14.9	14.6
Addendum						
Grants to State and Local Governments	4.7	7.6	9.2	12.2	15.3	15.8

SOURCE: U.S. Office of Management and Budget (1980a).

fulfill federal objectives such as stabilizing the economy. This objective was pursued most enthusiastically during the 1976-1977 economic stimulus program encompassing Local Public Works (LPW), Counter Cyclical Revenue Sharing (ARFA), and CETA Title VI programs.[3]

Along with the growth in grants have come two important policy shifts. First, an increasing fraction of federal assistance has been channelled directly to local governments, bypassing the states. Second, the recipients have been given a great deal more flexibility about how they spend their federal monies than was the case a decade ago.

As federal spending has grown and priorities have shifted, the ability of the Congress or the executive to control the budget from year to year has declined (see Table 1.4). Spending is considered "relatively uncontrollable" if it cannot be increased or decreased during a given fiscal year without changes in existing substantive law (U.S. Comptroller General, 1975). Much of the growth in uncontrollability stems from expanding entitlement programs—programs obligating the federal government to pay individuals or governments benefits if they meet certain eligibility requirements. Social security, food stamps, civilian and military retirement pay, unemployment insurance, Medicare, Medicaid, child nutrition assistance, welfare, and revenue sharing are the primary uncontrollable entitlement programs.

Interest, which absorbs $1 out of every $11 the federal government spends, is also uncontrollable, as are farm price supports and certain forms of student aid spending and housing assistance. Another category of "relatively uncontrollable" expenditures is the outlays from prior-year contracts and obligations. In this category are such things as the expenditures for procurement of a weapons system, the contract for which may have been signed with a private firm several years ago, and the ongoing payment for construction of a dam or public building that takes a number of years to build.

The spending for many of the programs regarded as relatively uncontrollable is driven by demographic and economic forces over which the Congress and the president have little control. For example, the rise in interest rates will cause interest

TABLE 1.4 Controllability of Budget Outlays, Fiscal Years 1967-1980

	1967	1970	1975	1980
Percentage Relatively Uncontrollable	59.5	64.0	72.8	76.1

SOURCE: U.S. Office of Management and Budget (1980a).

on the national debt to increase by about $7 billion between fiscal years 1979 and 1980. Spending on the unemployment insurance programs is expected to rise by $6.8 billion between fiscal years 1979 and 1980 because the recession will increase the number of persons unemployed. A similar impact will be felt in the food stamp, Medicaid, and welfare programs. Social Security and Medicare spending is driven partially by the growth in the population over age 65.

Not only are the number of persons applying for the benefits of entitlement programs beyond the immediate control of the government, but also the level of benefits is often geared to economic conditions rather than congressional decisions. The benefits of most entitlement programs are indexed—that is, adjusted automatically once or twice a year to compensate for rises in prices. The impact of this on the budget can be significant. For example, on an annual basis, the July 1980 automatic increase in social security benefits will add $16.3 billion to federal spending.

While the growing uncontrollability of the federal budget reduces the flexibility of policymakers, some of this loss of control is inevitable and some is even desirable. Uncontrollability associated with interest on the public debt falls into the inevitable category—only by a reduction in the public debt will this source of uncontrollability be reduced. The lack of control associated with entitlement programs falls into the second category. Congress could regain control over these spending programs by varying the benefits and eligibility rules of Social Security, unemployment insurance, food stamps, and similar programs each year depending on budget priorities and economic conditions. This, however, would remove the assurance citizens now have that public programs will protect them from extreme deprivation should they loose their means of support. If Social Security and Medicare benefits were the object of continual adjustment, rational planning for retirement would become impossible.

THE FEDERAL BUDGET IN THE 1980s

The course of federal budget policy during the next decade is no easier to forecast than the economy's. Nevertheless, several developments, certain to influence future federal budget policy, are apparent. These include the movements to limit spending and balance the federal budget, the increased importance accorded to national security and energy, and the growing impact of indexation.

The rapid growth in federal spending characterizing the past two decades is likely to be curbed during the next ten years. The support for this prediction comes from the pressure being exerted on the Congress and the executive branch to balance the budget and limit spending growth. This pressure is a manifestation of several general and spreading beliefs:

- that taxes are too high;
- that inflation is caused by governmental spending and deficits;
- that federal policies are encroaching on areas better left to private interests or state and local governments; and
- that federal programs have not worked and are fraught with waste and abuse.

The pressure has been expressed in several forms. Thirty state legislatures have petitioned the Congress either to pass a constitutional amendment requiring a balanced federal budget or to call a constitutional convention to take up such an amendment. Dozens of bills requiring a balanced budget have been introduced by senators and congressmen from both parties. The Senate Judiciary Committee refused by a 9 to 8 vote to report out a constitutional amendment for consideration by the full Senate in March 1980, but the issue is likely to be raised again.

While most observers feel that a constitutional amendment would be too drastic and inflexible a solution, there is a general acceptance that a problem exists and that the congressional and executive institutions are incapable of curbing the growth in spending or balancing the budget. Spending limitations—laws limiting federal spending—have been proposed as a less radical solution to the problem.

Most of the spending limitation proposals before the Congress follow one of three basic approaches. First, there are those that would set a maximum percentage rate of growth for federal outlays. An example of this approach is H.R. 867, introduced by Representative Wylie (R-Ohio), that would limit the growth of outlays and budget authority to 10% per year. Actions that would cause these limits to be breached could be considered only in time of war or by the passage of a concurrent resolution approved by a two-thirds vote of the total membership of both houses.

A second approach would limit federal outlays to a fixed percentage of some economic indicator. H.R. 5371, proposed by Representative Jones (D-Okla.), would limit federal outlays to 21% of the GNP in the first year and 20% thereafter. The limit could be broken through a presidential request for a waiver passed by a simple majority of both houses.

One variant on this approach is H.R. 6021, introduced by Chairman Giaimo (D-Conn.) of the House Budget Committee. To prevent the use of tax expenditures to circumvent the spending ceiling, this bill would limit the sum of direct outlays and tax expenditures to 28.5% of the GNP in the first year and 27.5% in the third and subsequent years. Another variant, H.R. 6706, introduced by Representative Dodd (D-Conn.), would limit spending first to 20% of the potential GNP and then to 19%. Potential GNP is the level of output achieved if the economy were fully utilizing its productive resources. This proposal would eliminate the problem of having the spending limit fluxuate perversely with economic cycles—a problem that could be severe because federal spending on programs such as unemployment insurance and food stamps tends to increase when economic growth slackens.

A third approach is to limit the rate of growth of federal spending to the growth of an economic indicator. House Joint Resolution 42 is a constitutional amendment, introduced by

Representative Guyer (R-Ohio), limiting the growth of budget authority and outlays to the average growth rate of the GNP in the three preceding years. An interesting variant of this approach has been proposed by the National Tax Limitation Committee and is embodied in H.J. Res 395 introduced by Representative Jenkins (D-Ga.). Under this proposal, spending could not increase at a faster rate than the growth of the GNP minus one-quarter of the percentage by which the inflation rate exceeded 3%. Thus, if the inflation rate were 11% and the GNP grew by 14%, federal spending could increase by no more than 12% (14% minus one quarter of the difference between 11% and 3%).

As currently written, most of the spending limitation bills would require drastic changes in projected federal budget policy. By and large, they would not permit spending at the levels needed to maintain existing programs with adjustments for inflation plus the defense policies implied by the budget resolution for 1981. Not only would existing programs have to be cut back, but no new spending programs could be passed by the Congress unless further reductions were made in current programs.

The impetus behind a balanced budget constitutional amendment and the spending limitations bills waned in the late spring of 1980 because of the administration's and Congress's determination to balance the fiscal year 1981 budget. In June, after considerable deliberations, the Congress passed the First Concurrent Resolution on the Budget Fiscal Year 1981, which called for a small surplus of $200 million. This resolution limited spending in many functional areas to increases that were substantially below levels needed to maintain current service levels in an inflationary environment. The resolution incorporated the first use of the reconciliation provisions of the congressional budget process; it instructed eight House and nine Senate authorizing committees to report legislation achieving savings in ongoing federal programs in fiscal year 1981.

The rapid deterioration of the economy during the first half of 1980 dashed hopes for a balanced budget in fiscal year 1981. Compared to the targets established by the first resolution on the 1981 budget, the recession lowered revenues by $9 to $14 billion and pushed up spending for such cyclically sensitive programs as unemployment insurance, trade adjustment assistance, and public assistance by $8 to $10 billion. Even without a tax cut or passage of spending programs designed to stimulate the economy, the fiscal year 1981 budget is likely to end up with a large deficit. This will occur in spite of the very real economies made by Congress and the president. This turn of events, combined with the realization that spending during fiscal year 1980 rose by over 17% and the deficit was close to the record level of $66 billion incurred in fiscal year 1976, should rekindle the movement to adopt some sort of spending limitation or balanced budget amendment. Even without the discipline of a spending limitation law, the spectre of such a bill passing is likely to hold down future growth in the budget.

The quarter-of-a-century shift in federal priorities from national security toward human resources should also come to an end in the next decade. Defense outlays as a fraction of all budget outlays have been stable at about 23% for three years. But budget decisions made in the past few years will cause defense outlays to rise at a rapid pace during the first half of the 1980s. Budget authority, an indicator of future defense spending, rose by 12.8% between fiscal years 1979 and 1980; this was over double the 6.1% average growth rate of the period 1970-1979. The budget resolution for 1981 calls for an increase of 20.2% for the coming fiscal year and for increases that could exceed 20% during fiscal years 1982 and 1983. The mood in the Congress and in the nation as a whole seems to favor an even greater increase in defense spending. Pay increases to maintain the quality and strength of the volunteer armed forces, new weapons systems, and the rapid escalation in the unit prices of many existing weapons systems make increases of these orders of magnitude probable.

Energy also will be an area of increased federal priority. Massive efforts to stimulate synthetic fuels production, conservation incentives, and the buildup of the strategic petroleum reserve will absorb budget resources that in the past decades flowed into human resources programs and grants to state and local governments. Energy-related outlays are projected to grow by 78% over the period 1980-1983, while related tax expenditures will increase by 83%.

Neither defense nor energy decisions are easily turned off, slowed down, or reversed. Budget decisions made this year and next will shape spending for the next five to ten years for major weapons systems (the MX missile, Trident submarine, B-1 bomber, and so forth), and synthetic fuel plants take years to plan and construct before they are put into operation.

Indexation is a final factor. In 1969, few federal programs had their benefits automatically indexed for inflation. In the course of the 1970s, this changed, until now more than 46% of federal spending is directly or indirectly indexed. Among the programs with benefits that are directly increased when prices change are Social Security, railroad retirement, civilian and military retirement, child nutrition, food stamps, veterans' pensions, and SSI. Medicare, Medicaid, AFDC, and unemployment insurance benefits are increased indirectly when prices rise.

The purpose of indexing is to protect the purchasing power or real services of the programs; but indexing in a period of rapid inflation tends to remove much of the government's budget flexibility. Roughly $34 billion, or 65.8% of the anticipated increase in federal spending between fiscal years 1980 and 1981 will be attributable to automatic inflation adjustments. This implies that during the 1980s there will be a further deterioration in the year-to-year control the Congress and executive branch can extend over federal spending. It also means there will be less room for discretionary programs—a category made up largely of grants-in-aid programs.

IMPLICATIONS FOR THE STATE AND LOCAL SECTOR

The state and local sector will be affected profoundly if certain aspects of the economic and budget forecasts sketched in the previous sections come to pass. The inflationary situation, the weak economic growth, the efforts to curb federal spending, and the shift in budget priorities will be particularly influential.

30 URBAN GOVERNMENT FINANCE

INFLATION

The spending and taxing systems of state and local governments, by and large, are not well adapted to an environment in which inflation averages over 7% and in which years of double-digit inflation are not extraordinary. Local government revenue systems, which depend heavily on property taxes and fixed fees, will be particularly hard hit. Most local governments have difficulty in rapidly reassessing real property, and taxpayer resentment over a nominal increase in assessment has been well demonstrated. Moreover, in an era of rapidly rising property values, the value of a home may bear little relation to a long-time owner's income or ability to pay the property tax. Inequities are therefore likely to arise with increasing frequency.

While state governments, with their sales and income taxes, should be in better shape, they too depend on revenue sources that do not always automatically keep pace with inflation—the gasoline tax being the most notable example. It is likely that the incompatibility of local revenue structures with an inflationary environment will cause a shift of revenue-raising responsibility and control to states, and away from local governments, during the 1980s. This has already begun in California and several other states that have placed limits on the ability of local governments to raise revenues.

State and local pension systems also could be hard hit by inflation. Those systems not indexing benefits will feel considerable pressure from their retired employees to raise benefits: several years of double-digit inflation are sufficient to turn an adequate retirement income into a pittance. Systems with indexed benefits may find their already shaky financial situations even more precarious as their liabilities unexpectedly mushroom. In general, underfunded pension systems should find their burdens increased as large numbers of workers retire at salaries blown-up by rapid inflation and hence are eligible for larger pensions than anyone previously estimated.

High inflation will mean a persistence of high interest rates. For those older jurisdictions facing a major rebuilding of their infrastructure, this will mean high borrowing costs and a reduced ability to undertake reconstruction. For younger, expanding areas, needed new infrastructure and capital will be more expensive.

Persistent inflation will also make determing wages in the public sector more difficult. For many in the economy, real wages will fall; but for those public employees with political power, job tenure, and no clear market test for their product, the focus is likely to be on the wage increases won by powerful unions in the private sector. Attention will be focused on "just" or "fair" pay increases—which will be defined as at least keeping up with the CPI—rather than on market wages. This will pose a problem as long as the CPI tends to overestimate the increase in the cost of living.[3] Tax increases sufficient to meet such wage demands will be difficult for many jurisdictions.

ECONOMIC GROWTH

The slow economic growth and high unemployment predicted for the 1980s will generate persistent demands for federal action to stimulate the economy. During the decade of the 1970s, such actions were a boon for the state and local sector. In some years, the federal antirecession spending programs—CETA Title VI, LPW, and ARFA—pumped more than $9 billion to these governments. By fiscal year 1980, outlays for these programs had dwindled to less than $2.3 billion, and, despite the worsening economic situation, fiscal year 1981 spending is projected to be even lower.

But federal fiscal policy is not likely to follow the path of spending increases and larger grants-in-aid again. For a number of reasons, stimulus is likely to take the form of tax cuts. First, there is considerable doubt concerning the true stimulative effect of past antirecession grants to state and local governments. Many observers believe that the Congress did not authorize and fund these programs until well after the economy had started to rebound; hence, the programs exacerbated inflation. Others feel that recipient governments spent the money too slowly or used it to substitute for local funds (Reischauer, 1979; Gramlich, 1979). Second, with spending limitations and balanced budget initiatives threatening, the Congress will be reluctant to engage in new programs that will increase spending as a fraction of GNP; tax cuts avoid this problem.

Third, the tax burden is reaching an all-time high. Without a tax cut, federal receipts as a percentage of the GNP will be over 22% in fiscal year 1981. This is higher than at any time in U.S. history. With disposable income per worker falling because of persistent inflation despite the 1980 recession, the pressure for tax relief will be overwhelming.[4] This pressure will recur throughout the 1980s. Rapid inflation will raise nominal incomes, thereby pushing taxpayers into higher marginal tax brackets. If a worker's wages keep pace with inflation, his after-tax income will fall behind because of the progressive nature of the personal income tax system. This effect will raise personal income tax receipts by almost $15 billion in fiscal year 1981 and has generated considerable support for indexing the federal income tax for inflation. Automatic inflation-related reductions in tax rates and in other aspects of the tax code have been adopted by Canada, Australia, Denmark, France, and nine other countries (Congressional Budget Office, forthcoming).

Finally, stimulative fiscal policies during the 1980s are likely to take the form of tax cuts rather than spending increases, because other important objectives can be pursued through the revenue system. One of these, of course, is the encouragement of investment and capital formation. Increased capital formation is essential if long-run economic growth and productivity growth are to be encouraged. A more rapid pace of productivity growth would reduce inflationary pressures. Liberalized depreciation rules, investment tax credits, or corporate income tax rate reductions are likely to be adopted to attain this goal.

Price stability can also be pursued through tax cuts that affect personal, as opposed to corporate, incomes. Social Security tax cuts or income tax credits for payroll tax payments are one such policy prescription. Such cuts would result in lower business costs, a portion of which should be passed forward to consumers in the form of reduced prices. And to the extent that personal tax cuts increase take-home pay, it may moderate future wage demands. This possibly could be enhanced by a tax-based incomes policy—whereby tax reductions would be limited to workers in firms that held down wage increases.

This scenario suggests that state and local governments will not be able to count on the federal government for significant fiscal assistance should they suffer the repercussions of a weak national economy. Large cities may be hit especially hard. The urban policy of the 1970s was a package of antirecession policies targeted on cyclically depressed, but more importantly, structurally sick, jurisdictions. As the political appeal of fighting recessions with increased federal spending wanes, this policy thrust could well disappear. To make matters worse, there is a possibility that the personal tax cuts that do occur in the 1980s will bypass those state and local workers (and federal employees) who are not currently paying social security taxes. This could lead to considerable bitterness and demands by public employees for higher wage increases. Investment-related tax cuts are likely to have unequal regional effects. New plant and equipment will be disproportionately located in the growing areas of the sun belt and in nonmetropolitan areas. The decline of the large depressed central cities in the Northeast and North Central regions is therefore not likely to be slowed by such policies (Northeast-Midwest Institute, 1978).

SPENDING LIMITATIONS AND SHIFTING PRIORITIES

The threat of spending curbs and the shift in priorities toward defense and energy are likely to result in a rapid decline in the importance of federal aid to state and local governments. The bulk of the federal budget consists of mandatory entitlement programs, such as Social Security and defense spending. Most of the rest—the controllable portion of the budget not defense or energy—consists of grants-in-aid to state and local governments. It is from this segment that budget reductions, forced or encouraged by spending limitations, are likely to come.

In a relative sense, such reductions are already taking place. As a fraction of the federal budget and of state and local expenditures, grants-in-aid peaked in 1978 (see Table 1.5). Congressional actions for fiscal year 1981 call for a continuation of this trend. Grants will increase significantly slower than both the inflation rate and the growth in total federal spending.

TABLE 1.5 Federal Grants-in-Aid As a Percentage of Federal Outlays and State and Local Expenditures, Fiscal Years 1960-1980

	Federal Outlays	State and Local Expenditures
1950	5.3	10.4
1960	7.6	14.7
1970	12.2	19.4
1975	15.3	23.1
1978	17.3	26.4
1979	16.8	25.6
1980	15.8	25.3

SOURCE: U.S. Office of Management and Budget (1980a).

Grants did not fare well in the budget-cutting exercise accompanying the formulation of the first budget resolution for fiscal year 1981. The state government share of revenue sharing, antirecessional fiscal assistance, and LEAA grants were eliminated; CETA Title VI jobs were reduced in spite of the rising unemployment rate; and EPA and highway construction grants and UDAG and CDBG programs were reduced. Funding for numerous other grant programs was held constant or permitted only modest increases. The continuation of a constant funding level and double-digit inflation can soon transform a substantial grant program into an empty shell. The general revenue sharing program is an example. In spite of periodic increases in funding, the real value of this program has been reduced by over 30% between 1974 and 1980 because of inflation.

To make matters worse, the political coalition necessary to defend the relative slice of the budget garnered by state and local governments could weaken. As long as the pie was expanding and there was a commonality of interests and problems, major types of state and local governments from the different regions worked well together as an effective lobbying force. As the competition for federal money becomes more intense and the problems experienced by the different regions and types of governments become less similar, this coalition could splinter.

The economic conditions faced by the various regions are likely to be very different in the 1980s (U.S. Department of Commerce News (1980). The current economic downturn is

TABLE 1.6 State Revenues Resulting from the Phased Decontrol of Domestic Oil Prices; 1980-1990

	Total Revenue (billions of dollars)	Average Annual Revenue (billions of dollars)	Annual Average Revenue as a Percentage of 1978 Own-Source Revenue
Texas	33.2	3.0	44.5
Alaska	37.3	3.4	398.5
Louisana	13.8	1.3	47.9
California	21.8	2.0	11.6
Oklahoma	3.1	0.3	16.5
Wyoming	3.5	0.3	84.4
New Mexico	1.5	0.1	12.5
Kansas	0.9	0.1	6.3
All Others	12.2	1.1	1.1

SOURCE: U.S. Treasury Department.

expected to leave much of New England, the South, Southwest, and West relatively unscathed. Increased defense and energy spending are likely to have the same kinds of unequal regional effects. This was already evident in May 1980. While the national unadjusted unemployment rate stood at 7%, that in Michigan was 14%, Indiana 10.5%, and Ohio 8.7%. At the other extreme were Texas at 5.0%, Florida at 5.3%, and Massachusetts at 5.8%.

The rise in energy prices will provide a bonanza to a handful of these same states. As domestic oil prices increase, severance tax collections, royalties, and earnings from state-owned oil properties will rise. One estimate of the oil receipts likely to occur to specific states is provided in Table 1.6. It should be noted that sales of oil from lands owned by states and localities are exempt from the new windfall profits tax. Coal prices also will increase rapidly, sending severance tax receipts from coal up sharply. While states and localities will experience added public service burdens from expanding energy production, these revenue demands will be well below the added receipts generated by such development (Congressional Budget Office, 1980).

Most of the increase in energy-related revenues will be paid in the form of higher prices by consumers located in nonproducing states.[5] These new revenues will allow some fortunate states to increase services, cut taxes, and attract business to a degree unprecedented in the last decade. Thus a significant segment of the state and local sector may well be able to bear a sharp decline in the real value of federal assistance. An extreme example of this is occurring in Alaska, which has used its oil revenues to abolish its personal income tax and to pay tax refunds to long-time residents of the state.

The same areas that will have relatively strong economies will be gaining in political strength. Preliminary projections suggest that the congressional reapportionment of 1982 should result in losses for the Northeast and North Central areas and gains for the South and West. New York could lose five congressional seats; Ohio, Pennsylvania, and Illinois could lose two each; and Indiana, Massachusetts, Michigan, Missouri, New Jersey, and South Dakota could lose one each. The projections suggest that Florida could gain four seats, Texas three seats, and California two seats; and Arizona, Colorado, Nevada, New Mexico, Oregon, Tennessee, Utah, and Washington one seat each. Public sentiment in the states gaining representation has generally been less supportive of expanded governmental activities and more favorable to states rights.

In addition to the interstate shifts, intrastate reapportionment should affect the future of urban policy. Recent central city and metropolitan area population losses will be reflected in a loss of congressional representation for big city interests.

Economic, budgetary, and political forces are converging in a way that will rend the expectations concerning fiscal federalism that have been built up over the past two decades of ever-expanding federal financial involvement. Federal grants-in-aid are likely to dwindle in importance during the 1980s. Specific programs designed to deal with the problems of older cities will probably fare the worst.

NOTES

1. The direct effect was 2.3 percentage points. The indirect effect was 0.3 percentage points and is due to oil decontrol which calls for the price of domestically produced oil to rise gradually to world levels. Over the decade, oil prices rose by 30% per year; in the 1960s, they declined slightly.

2. Such an investment would be inefficient from an economic perspective as long as the cost of producing synthetic fuels was above the cost of producing additional

supplies of conventional fuels. From a national security or risk-of-minimizing perspective, such an investment could be efficient.
3. The CPI can be a misleading measure of changes in the cost of living for several reasons. First, it is a fixed-weight index that reflects consumption patterns of the 1972-1973 period. Consumers substitute away products whose prices have risen most rapidly—such substitution is not reflected in the CPI. Thus the cutbacks in energy use brought on by the relative price increases of gas and oil are not reflected in the CPI. Second, the method of incorporating homeownership costs into the CPI has been criticized for ignoring that a home is an asset as well as a source of shelter and for overemphasizing the importance of changes in the mortgage interest rates. The personal consumption expenditure defaltor (PCE), another government-issued cost-of-living index, suffers from neither of these flaws. In the 12 months ending in May 1980, the PCE increased by 10.6% while the CPI rose by 14.4%. For a discussion of these issues see Congressional Budget Office (1979).
4. Despite slow economic growth and high rates of inflation, real disposable income grew at a fairly rapid pace during the 1970s. Much of this growth came from the increase in the labor force, which can be seen from the following table.

AVERAGE ANNUAL GROWTH OF REAL DISPOSABLE PERSONAL INCOME

	Total	Per Capita	Per Worker
1950-54	11.1	3.8	8.4
1955-59	12.2	4.6	6.5
1960-64	18.5	11.6	12.9
1965-69	16.3	11.5	7.3
1970-74	13.5	9.8	3.2
1975-79	15.7	12.1	4.1

The expected slowdown in the rate of growth of the labor force will reduce this source of growth in real income.
5. Unless the current distribution formula is changed, increased severance tax receipts will add to the tax-effort factor in the revenue-sharing formula, thus allocating these state areas larger payments.

REFERENCES

CRANDALL, R. W. (1978) "Federal Government initiatives to reduce the price level." Brookings Papers on Economic Activities 2.
Data Resources Inc. (1980) U.S. Long-Term Review (Summer).
DENISON, E. F. (1979) Accounting for Slower Economic Growth: The United States in the 1970s. Washington, DC: Brookings Institution.
GRAMLICH, E. M. (1979) "State and local budget surpluses and the effect of federal macroeconomic policies," U.S., Congress, Joint Economic Committee, 12 June. Washington: Government Printing Office.
Northeast-Midwest Institute (1978) "Investment and employment tax credits: an assessment of geographically sensitive alternatives." June 2.

REISCHAUER, R. D. (1979) "Federal countercyclical policy—the state and local role" in National Tax Association—Tax Institute of America, Proceedings of the Seventy-First Annual Conference, Columbus, Ohio.

U.S. Congress, Congressional Budget Office (1980) Indexing the Individual Income Tax for Inflation. Washington: Government Printing Office.

——— (1979) Entering the 1980s: Fiscal Policy Choices. Washington: Government Printing Office.

——— (1980) Energy Development, Local Growth, and the Federal Role. Washington: Government Printing Office.

U.S. Comptroller General of the United States (1975) Budgetary Definitions. Washington: Government Printing Office.

U.S. Department of Commerce (1980) "Sensitivity of states to the national business cycle." U.S. Department of Commerce News (15 June).

U.S. Office of Management and Budget (1980a) Budget of the United States: Fiscal Year, 1981. Washington: Government Printing Office.

——— (1980b) Federal Government Finances. Washington: Government Printing Office.

WACHTER, M. L. (1976a) "The changing cyclical responsiveness of wage inflation." Brookings Papers on Economic Activity No. 1.

——— (1976b) "The demographic impact on unemployment: past experience and the outlook for the future," in National Commission for Manpower Policy (ed.) Demographic Trends and Full Employment, Special Report 2 (December).

2

Public Employment's Impact on the Future of Urban Economies

SHAWNA GROSSKOPF
Southern Illinois University at Carbondale

☐ IN 1950, 10.6% OF EMPLOYMENT in urban areas was in the public sector. By 1970, it had increased to 16.5%. Obviously, public employees have had and will continue to have an important effect on urban economies. Not only are they a large portion of the urban labor force, but their employment and compensation have an important effect on the fiscal position of urban governments. Public employees also have a relatively unique effect on the quality and level of services delivered; both directly through their productivity on the job and indirectly through their voting power.[1] Thus the future of public employment will strongly affect the future of urban economies.

The received wisdom known as Wagner's Law (1877) suggests that the public sector has grown and will continue to grow as a consequence of an industrialized society shifting its demand from private to public goods. One can add to this backdrop relatively recent evidence that: (1) The demand for local public employees is relatively inelastic (see Ehrenberg, 1973); (2) that unit costs are rising in the public sector (see Bradford et al., 1969); and (3) that those costs could theoretically rise without

AUTHOR'S NOTE: *The author would like to thank Jesse Burkhead, who initiated and cowrote earlier work on this topic; and Claudia Striegel, for her contribution in typing many drafts of the manuscript.*

40 URBAN GOVERNMENT FINANCE

limit (see Baumol, 1967). If unit costs continue to rise because of increasing wages of public employees as suggested in (2) and rise without limit as suggested by (3), the fact that the demand for local public employees has been found to be inelastic (see 1) implies that local government expenditures will spiral, apparently without control.

Yet local governments *must* control their expenditures. Not only does a stagflation economy adversely affect the revenue-raising ability of local governments, but the increasing prevalence of state mandates also limits their spending levels. Local governments have traditionally relied on the property tax to generate most of their revenue (although the income and sales taxes are becoming more common) (ACIR, 1978). During the last two decades, urban governments (especially those outside the sunbelt) have been losing nonresidential components of their tax base as firms have moved to the suburbs or to the south to be followed (or preceded) by residents. The relatively low income groups remaining in urban areas, the deteriorating housing stock, spiralling interest rates, and falling federal aid paint a bleak picture for the fiscal future of urban governments.

The economic patterns of the recent past, however, have been those of profound structural change. It is not clear that conventional wisdom about the growth of the public sector applies to such periods, nor that the 1980s will see a return to the patterns of growth of the 1950s and 1960s. More appropriate are the trends observed in the 1970s. Also appropriate is a critical look at the received wisdom in light of the experience of the 1970s. Accordingly, this chapter analyzes the recent trends in public employment and compensation. Throughout, an attempt will be made to assess this recent evidence against the received wisdom. The last section summarizes the evidence to gain some insight into the problems and prospects facing urban governments in the 1980s. These trends should shed some light on the prospects of urban governments controlling their expenditures for personnel. If those expenditures are uncontrollable during a recessionary period such as the 1970s, it is almost unavoidable that more and more urban governments will face fiscal and political crisis in the 1980s.

TRENDS IN PUBLIC EMPLOYMENT

There are two classic references suggesting that the public sector will grow faster than the private sector. Wagner's Law is basically a demand formulation of the growing public sector hypothesis: As incomes grow, demand shifts toward public goods. A simplified explanation of this model is that the demand for public services is relatively income elastic. If public sector output grows proportional with public employment, then Wagner's model implies growing public employment, at least during periods of economic growth.

In contrast to this strictly demand side approach, Baumol (1967) employed a productivity or supply-side approach to explain growth in public employment. In his model of unbalanced growth, he assumes that the economy can be divided into broad sectors: (1) a progressive sector characterized by increasing productivity; and (2) a nonprogressive sector characterized by constant productivity. Interpreting government as a nonprogressive sector, Baumol's first proposition implies that the relative cost of public to private goods will rise over time (without limit). If we make the further assumption that the demand for public goods is price inelastic, Baumol's proposition 3 implies that the public sector will require ever increasing amounts of employees to maintain its production, displacing employment in the private sector.

Another growing strand of thought suggests that public employees are an important source of the recent growth in the public sector. Recent work by Borcherding (1977) suggests that only 50% of the growth in public sector spending between 1902 and 1970 can be explained by economic arguments such as: increase in expenditures due to inelastic demand and rising prices or increases in expenditure due to rising incomes (i.e., Wagner's law). He concludes that at least half of the growth in the public sector could be due to political pressures. One which he emphasizes is the political or monopsony power of public employees themselves.

Not only is there evidence that a larger percentage of public than private employees vote in local elections (Bush and Denzau, 1977), but they also represent a potential interest group with incentives to increase public spending either through a perceived "lower" price of services due to increased wages (Borcherding et al., 1977) or through a desire to maximize their own power and budgets (Niskanen, 1971). Coupled with fiscal illusion and an inelastic demands for public employees, there is (according to this public choice view) little likelihood that public employment will stop growing.

This particular approach ignores the inherent limitations of the monopsony power of public employees. Recent work by Courant et al. (1979) presents convincing theoretical evidence of the limitations of public employee power. Given that private workers in a given jurisdiction are not immobile, the potential threat of a falling tax base will limit attempts by public employees to increase wages and employment in the public sector, at least at the local level.

It should also be noted that none of the three models above discusses the implications of adverse economic conditions on the predictions of their models. Both the Baumol and Wagner models assume that the economy in question is expanding, i.e., there is growth in real GNP.[2] The public choice approach, rooted firmly in positive analysis, has had little chance to analyze governmental behavior in an economic framework of decline and slow growth exhibited in the 1970s and predicted for the 1980s. Given that urban economies are classic examples of open economies, it is clear that a national economy characterized by inflation *and* recession will not have a neutral impact on urban economies and their local governments.

Thus the received wisdom suggests that public employment should continue to grow, yet that wisdom was based implicitly on the assumption that real incomes are rising—an assumption not tenable for the 1970s and will be unlikely to hold consistently in the 1980s.

In this light, the evidence on public employment trends in the 1970s is critical in determining the implications of these models for the 1980's. Accordingly, employment trends for the post-World War II period are presented, with emphasis on the behavior of employment in the last decade.

The patterns of the longer past should not, however, be ignored. They will be used to gain some insight into the patterns of public employment over the business cycle, which in turn should clarify the applicability of the theories cited above in predicting public sector activity as the economy fluctuates. This cyclical stability of public employment will be discussed as a separate topic after the discussion of recent employment patterns.

RECENT EMPLOYMENT TRENDS

With about 1 out of every 6 workers in the United States employed by the public sector in 1978, public employees are a significant part of the labor pool. Local government employment alone accounts for almost 1 out of every 10 employees. These statistics reflect a significant increase since World War II in public employment as a share of total employment. In 1948, 1 in 20 employees worked for local governments, with 1 in 10 workers in the public sector as a whole.

The data on employment trends are compiled from the Census of Governments and the Survey of Current Business. Table 2.1 refers to employment totals by sector, where employment refers to the number of full-time and part-time employees for the period 1948-1978 (Table 2.1). Full-time equivalent employment was consistently available only after 1963 (Table 2.1).

Most of the increase in the public sector share of total employment occurred between 1948 and 1968 (Table 2.1). Since 1968, however, the public sector share has remained almost constant. In contrast, service sector employment as a share of total employment has increased steadily over the entire thirty-year period.[3]

Service sector trends are of interest in this case for two reasons. Service and public employees are engaged in producing services rather than goods, and so it is difficult to quantify what they produce. Thus service workers are a private sector counterpart to public employees, and this will be especially useful when discussing compensation. Related to this similarity, service employment could therefore also be included in Baumol's nonproductive sector. Thus, using the public and service sector total, nonproductive employment has increased steadily from

TABLE 2.1 Employment Trends by Sector, 1948-1978

	Employment by Sector as a Percentage of Total Employment (Full-time Equivalent Employees)				
Year	All Governments	Federal Civilian	State	Local	Services
1978	15.3%	2.4%	3.6%	9.3%	18.2%
1977	15.8	2.5	3.7	9.7	18.0
1976	15.9	2.6	3.7	9.7	17.7
1975	16.2	2.7	3.7	9.9	17.5
1974	15.4	2.6	3.5	9.4	17.0
1973	15.2	2.5	3.4	9.3	16.7

SOURCES: U.S. Bureau of the Census (1979a, 1979b), U.S. Department of Commerce, Bureau of Economic Analysis (1976).
NOTE: Totals may not sum exactly, due to rounding. Services refer to the industrial classification used by the Census.

about 20% of total employment in 1948 to more than 35% in 1978.

For growth rates in employment (Table 2.1), services and government grew more rapidly than total employment in 1948-1972. While employment grew rapidly during that period, the most dramatic increases were in state and local government employment.

To get a more accurate idea of total employment behavior during the 1970s, we will henceforth rely on employment trends based on full-time equivalent employees (FTE) [Table 2.3]. This will eliminate the problems of comparability that would arise if the various sectors considered here do not have similar shares of part-time employees. If adjustment in one sector is based entirely on hiring and firing part-time employees, employment changes would be overstated, relative to sectors relying more heavily on full-time employees.

Until 1973, as employment grew overall, the sectoral composition of employment in the United States shifted toward a more public dominant labor force, consistent with Wagner's law and Baumol's model. This simple pattern of growth in sectoral employment did not, however, continue through the rest of the decade.

In the period 1973-1974, when the shock of the oil embargo was first felt, growth in governmental employment exceeded

TABLE 2.2 Average Annual Growth in Employment[a] (Full-time Equivalent Employment)

Year	Total Employment	All Governments	Federal Civilian	State	Local	Services[b]
1977-78	4.60%	1.27%	1.35%	2.17%	0.91%	5.58%
1976-77	3.58	3.16	0.05	3.72	3.79	5.33
1975-76	3.16	0.99	0.61	2.00	0.72	4.56
1974-75	-2.75	2.31	1.38	3.43	2.15	0.23
1973-74	1.47	2.82	2.62	4.16	2.39	2.91
1973-78	2.93	2.11	1.20	3.10	1.99	3.72
1968-73	1.54	3.02	-5.60	4.10	3.95	3.06
1963-68	3.52	5.53	2.75	5.60	4.18	4.87

SOURCES: See Table 2.1.

[a]Derived from compounding formula: $i = \left[\left(\frac{FV}{PV}\right)^{1/t} - 1\right] \times 100$, where FV is future value, PV is present value, and t is the tth period.

[b]Services refer to the industrial classification used by the Census.

TABLE 2.3 Average Annual Growth in Employment[a] (full-time and part-time employees)

Years	Total Employment	All Governments	Federal Civilian	State	Local	Services[b]
1977-78	4.23%	1.11%	1.40%	1.37%	-0.75%	4.91%
1976-77	3.50	3.00	0.18	4.43	5.08	5.21
1975-76	3.19	0.26	-1.63	2.20	0.15	3.87
1974-75	-1.34	2.36	0.56	3.68	2.49	2.55
1973-74	1.81	3.46	3.16	4.71	3.12	2.37
1973-1978	2.28	2.04	0.73	3.28	2.02	3.78
1968-1972	2.15	2.76	1.35	3.85	3.97	2.93
1963-1968	2.30	4.87	3.72	7.06	4.87	3.85
1958-1963	1.48	3.25	1.27	4.75	3.84	3.32
1953-1958	1.97	3.32	0.93	5.42	4.62	3.68
1948-1953	1.44	3.47	6.80	2.77	3.74	2.42

SOURCES: See Table 2.1.

[a] Derived from compounding formula: $i = \left[\left(\frac{FV}{PV}\right)^{1/t} - 1\right] \times 100$, where FV is future value, PV is present value, and t is the tth period.

[b] Services refer to the industrial classification used by the Census.

the total growth but not service employment. During the 1974-1975 recession, total FTE employment actually contracted and service employment growth was at a virtual standstill. But governmental employment continued to increase. Thus, during the trough of the general employment contraction, governmental employment (especially state and local employment) was stabilizing. But as the rest of the economy began to recover after 1975, governmental employment grew slowly, contributing little to the recovery. In fact, with the exception of 1976-1977, governmental employment trends indicate a gradual decline in the rate of increase over the period 1973-1978.

These trends indicate that services increased their share of employment during periods of growth and decline; i.e., the rate of growth in service employment exceeded that of total employment in both periods of growth and decline. But government's share of total employment declined slightly after 1974, despite mildly countercyclical behavior during the recession of 1974-1975.

Thus, recent trends in state-local employment indicate an emerging countercyclical pattern. The boom period of 1950-1972, in which growth in state and local employment suggested a relative increase in public versus private goods provision in the economy was not repeated in the 1970s. During the recession of 1974-1975, state and local employment was mildly countercyclical. Although service levels (employment) did not grow at prerecession rates, they did exceed those of total employment. After 1975, as real personal income began to increase, state and local employment growth lagged behind total employment growth. Employment levels declined relative to previous growth rates in state and local employment and relative to private sector growth rates.

Thus, recent trends in public employment do not conform to Wagner or Baumol's predictions. During the period of falling income, public employment grew. During the period of recovery and growing income, public employment growth rates lagged behind the rest of the economy. The 1970s were obviously not a period of simple growth nor are the 1980's likely to be. Based on the recent trends in public employment, it appears

48 URBAN GOVERNMENT FINANCE

that public employment growth may level off, at least during the early 1980s.

EMPLOYMENT STABILITY

Employment patterns in the public sector were relatively erratic in the 1970s. We now turn to a longer view to determine what cyclical patterns have emerged in public and service employment over the last 30 years. This should capture the resiliency of employment in these sectors to changes in economic activity over the business cycle.

Due to the employment shifts in the 1970s, the cyclical stability of these sectors is of particular importance for urban areas. As manufacturing left urban areas in the 1970s, the service and public sectors emerged as a major part of the new urban employment base. Whether or not this new nonindustrial economic base will provide an economic and financial foundation for urban economies remains open to question.[4] We address that issue by investigating the resiliency of this employment base to economic downturn based on data from the past 30 years. We are trying to measure the cyclical stability of public and service employment.

Previous work on the cyclical performance of the state and local sector concentrated on revenues and expenditures over the cycle. It was widely held that the behavior of state and local government was fiscally perverse, i.e., destabilizing. This was, in part, the result of a study by Hansen and Perloff who concluded that "these governmental units have followed the swing of the business cycle, from crest to trough, spending and building in prosperity periods and contracting their activities during depressions" (1944). More recent studies find that the postwar patterns are not consistently destabilizing.

Both Rafuse (1965) and Ross (1978) concur that the state and local sector is characterized by growth. They find that, during recessions, state and local revenue and expenditure behavior is countercyclical, i.e., stabilizing. They also find that there is a recent tendency for state and local governments to make rather weak contributions to recovery after recessions.

The issue is whether public employment is cyclical or countercyclical—providing indirect evidence on the stability of service levels over the business cycle. Ross (1978) includes a brief summary of public employment trends during expansions and contractions in the economy between 1957 and 1976. Based on growth in full-time equivalent employment, he finds that state and local employment increased during recessions, although the countercyclical effect decreased with each successive recession since 1957. He also finds progressively smaller increases in employment during periods of recovery. State and local full time equivalent employment is, thus, countercyclical during recessions.

The method used here to measure cyclical stability of employment is relatively simple. Since changes in employment levels over time include trend and cyclical patterns, the first step in determining the cyclical stability is to remove the trend component. This is accomplished by regressing employment (in public, total, and service sectors) on a time trend variable. The residuals from this regression (divided by the predicted level of employment) then are the cyclical patterns of employment, i.e., the deviations from trend over the period 1948-1978. This measure of the cyclical volatility of employment is then regressed on that for total employment. The coefficient of this regression measures the cyclical sensitivity of employment in the public and service sectors relative to cyclical changes in total employment.

This measure of cyclical sensitivity has a simple interpretation: If governmental employment fluctuates more than total employment, it is procyclical employment and will have a coefficient greater than one; if government employment fluctuates less than total employment, it is countercyclical employment with a coefficient of less than one.

The resulting "elasticities" are displayed in the first column of Table 2.4. Based on the period 1948-1978, local government employment is less stable (i.e., more procyclical) than total employment. Federal government employment, as would be expected, is relatively stable (countercyclical). Services and state government employment, however, are the most strongly cyclical.

This result does not seem to corroborate the results of Rafuse and Ross. This could be due, in part, to the difference in the

TABLE 2.4 Cyclical Employment Response Elasticities of Public and Service Employment, 1948-1978

Employment Sector	Full-Time and Part-Time Employment		
	Elasticity Coefficient	Intercept	R^2
All governments	1.5787	.0030	.7761*
Federal government	0.3135	.0126	.0305
State government	4.2763	.0362	.6899*
Local government	1.9488	.0061	.7182*
Service sector	2.8963	.0092	.8266*
	Full-Time Equivalent Employment		
	Elasticity Coefficient	Intercept	R^{2b}
Federal government	0.7277	-0.0005	0.2546*
State-Local government	1.0382	0.0070	0.3028*
Service sector	1.4035	0.0044	0.4812*

SOURCES: See Table 2.1.
NOTE: The elasticity coefficient is calculated from the following equation:

$$\left(\frac{E_i - \hat{E}_i}{\hat{E}_i}\right)_t = a_t + b_t \left(\frac{E_j - \hat{E}_j}{\hat{E}_j}\right)_t,$$

where E_i is employment in sector i in year t, $\hat{E}_i$ is predicted employment based on the preliminary estimating equation (see text). E_j and $\hat{E}_j$ are the corresponding variables for total employment. Thus, a is the intercept and b is the elasticity coefficient.
*Significant at the .05 level of confidence.

measures used to measure cyclical responsiveness. Another alternative is that the use of full-time and part-time employment as the measure of employment rather than full time equivalent employment implies an overstatement of the cyclical responsiveness of employment. If state and local governments react to cyclical changes by hiring or laying off part-time rather than full time employees, the variation in total employment would be larger than the variation in full time equivalent employment.

The elasticity coefficient for full time equivalent employment indicates that the response elasticities are smaller than those for full and part-time employment (except for the federal government). Based on FTE employment, the state and local elasticity dropped to roughly unity. The service employment elasticity remains, however, significantly greater than unity.

Overall, then, it appears that, although urban governments have been able to maintain public employment levels that are not destabilizing, they face increasing employment instability as service employment grows. This is particularly important for "older" cities for whom the service sector of their employment bases is growing most rapidly.[5]

In this context it is also important to note that the results in this chapter are based on total trends. At a more disaggregated level, it is very likely that public employment is contracting during recessions. It is even more likely that the general economic decline of the "older" cities, particularly those in the snowbelt, will cause them to contract public employment during declines, whereas the younger cities will continue to grow as employment shifts regionally, an issue meriting further study.

The evidence from the previous section does suggest that the long-term boom in public employment may be waning at least for the recent past and near future. This implies that service levels will not be growing at rates realized during the 1950s and 1960s, unless unexpected technological advances drastically increase productivity in state and local governments. Relatively slower growth in public employment could also imply more stable expenditure patterns.

Stabilization of employment does not, however, tell the whole story; there remains the issue of how public employee compensation has adjusted to the changing economic conditions during the recent past. Even if local governments can control employment growth, they may not succeed in controlling price increases in public employment. In that case, urban economies face stagant service levels with rising costs.

COMPENSATION

Compensating public employees is an important part of urban governmental budgets: Local government expenditures for personal services as a percentage of direct expenditure has remained at almost 50% from 1960 to 1978. Spiraling wage increases in the public sector, especially if demand for public employees is relatively inelastic as Enrenberg (1973) suggests, could have a critical effect on the fiscal health of urban governments in periods of stagflation. The degree to which urban governments can control public employee wage increases in the 1980s becomes a critical issue.

Before looking at recent trends in compensation of public employees, some effort should be made to analyze wage determination in the public sector. Although there is an extensive literature purporting to explain why, or empirically verify that, public compensation levels exceed those in the private sector, relatively little has been done to specify the cyclical behavior or determinants of government wages.[6]

Unfortunately, the neoclassical model of marginal productivity theory and wage determination is of little help in analyzing public sector wage determination, since marginal product has little meaning when the "good" being produced is a public good, and the employer is not a profit maximizer. Nonetheless, particularly at the local level of government, a governmental official who hopes to be reelected will have some incentive to minimize costs, implying "competitive" behavior in the factor market. In contrast, the pressures from well-informed constituents felt by local government officials diminishes at the state level and is least effective at the federal level. It is very unlikely that an individual knows the salary of a bureaucrat in HEW, for example, but they might be aware of the level of salaries of teachers in their own municipality.

There are those who argue that public sector wage determination is affected by noncompetitive characteristics of employment. Reder (1978) argues that public employees could receive pecuniary benefits from their role in increasing the votes of the bureaucrat. Bush and Denzau (1977), among others, argue that public employees, through their voting power, can influence the level of public expenditures and their own employment and wages. Again, this ability is limited by the mobility of other residents of the jurisdiction (Gramlich et al., 1979). The success of the taxpayers' revolt also suggests that there are limits to the ability of public employees to increase local government expenditures and employee wages.

Baumol's model of unbalanced growth, on the other hand, implies that wages in the nonproductive sector will be determined by those in the private or productive sector. Since marginal productivity is relatively stagnant in the governmental sector, wages there will eventually exceed marginal product; hardly a cost-minimizing result.

Baumol does not, however, specify what will happen if the economy is not growing. But it does seem unlikely that governmental wages would immediately follow if private sector wages fell, although there would be downward pressure on wages as workers shift from the private to public sector, creating excess supply.

If one takes the approach to Wagner's model that the demand for public goods is income elastic, then a decline in income implies a decline in public goods demand. Given a proportional relation between public employment and output and a less than perfectly elastic supply of labor to the public sector, a decline in national income would induce a fall in the demand for labor. This in turn implies a reduction in employment and wages. This, again, implicitly assumes competitive factor market behavior.

Thus, the theory does not unambiguously predict the pattern of wages in the public sector—that issue remains empirical. But before examining relative wage trends, several caveats are in order. Although the data prepared by the Bureau of Economic Analysis (Department of Commerce) are adequate for many purposes, there remain difficulties in defining compensation. Most problems lie in evaluating "fringes," i.e., pensions, employer contributions in the form of sick pay, medical and dental insurance schemes, group life insurance, vacations, subsidized lunch rooms, subsidized recreation facilities, and others. And compensation paid by the employer is not necessarily equivalent to compensation as viewed by the employee. For example, early retirement for policemen or firemen that facilitates a second career may be more valuable to the employee than the cost of the pension contributions by a local government.

There are further difficulties. If average skill levels are higher in one sector than the other, it would be reasonable to anticipate that compensation levels would be correspondingly higher. Examining aggregate data for public employees suggests that this sector may have more highly skilled employees, at least if one may assume that educational attainment corresponds with on-the-job skills—a reasonably heroic assumption. The 1970 Census reported that 59% of private sector employees had from one to four years of high school education, while only 43% of governmental workers (federal, state, local) had comparable educational attainment. But 45% of governmental employees had from one to five or more years of college, while only 21% of the private civilian labor force was in this category.

Keeping the difficulties of comparing public and private sector compensation in mind, the available data are summarized in Table 2.5. Until 1967, state-local average employee compensation lagged behind the private (total) sector (but ahead of services)—both for wages and salaries and for supplements (Table 2.6). By 1967, the differential was overcome—not by an increase in wages and salaries, which still lagged behind the private sector, but by increases in supplements. By 1972, total compensation and wages and salaries per full-time equivalent employee exceeded those of private industry and supplements lagged.

The turn-around came in 1975, when total compensation in the private sector edged ahead of the state and local sector. The rate of improvement in total compensation, which had consistently exceeded that in private industry before 1972, fell below the private rate in 1973 and continued to lag through 1978 (with the exception of 1975-1976). Supplements jumped sharply for state and local employees in 1974 increasing at faster rates than the private sector through 1978.

Comparisons with movements in the consumer price index (CPI) show much the same pattern (Table 2.7). There was consistent improvement in real income for state and local employees from 1962 to 1972, with wages, salaries, and supplements exceeding the increase in the CPI and exceeding the relative improvement in compensation in private industry; 1973 marks the turn. Total compensation increased at a rate less than the increase in the CPI; real income fell. Not until 1976 did real compensation actually increase for state and local employees, only to fall again in 1977-1978. Only supplements grew consistently in the 1970s in real terms.

TABLE 2.5 Annual Compensation, Wages and Salaries, and Supplement (per full-time equivalent employee), 1952-1978

Year	Total Compensation		Wages and Salaries		Supplements		Supplements as a Percentage of Wages and Salaries	
	State and Local	Total	State and Local	Total	State and Local	Total	State and Local	Total
1978	$15,585	$15,693	$12,966	$13,275	$2,619	$2,418	20.2%	18.2%
1972	14,543	14,558	12,245	12,382	2,298	2,176	18.8	17.6
1976	13,521	13,527	11,570	11,600	1,951	1,927	16.9	16.6
1975	12,498	12,519	10,841	10,835	1,656	1,684	15.3	15.5
1974	11,464	11,451	10,029	9,991	1,435	1,460	14.3	14.6
1973	10,729	10,584	9,482	9,300	1,247	1,284	13.2	13.8
1972	10,009	9,885	8,899	8,760	1,110	1,125	12.5	12.8
1967	7,007	6,962	6,284	6,307	723	655	11.5	10.4
1962	5,449	5,630	4,987	5,162	462	468	9.3	9.1
1957	4,282	4,608	3,958	4,299	324	309	8.2	7.2
1952	3,345	3,647	3,140	3,454	205	193	6.5	5.6

SOURCE: U.S. Department of Commerce, Office of Business Economics, *Survey of Current Business*, July issue (various years).
NOTE: Supplements equal total compensation (per full-time equivalent employee) less wages and salaries.

TABLE 2.6 Average Annual Growth Rates in Compensation, Wages and Salaries, and Supplements, 1972-1978 (per full-time equivalent employee)

Years	Total Compensation		Wages and Salaries		Supplements	
	State and Local	Total	State and Local	Total	State and Local	Total
1977-78	7.17%	7.80%	5.89%	7.21%	14.0%	11.1%
1976-77	7.56	7.62	5.83	6.74	17.8	12.9
1975-76	8.19	8.05	6.72	7.06	17.8	14.4
1974-75	9.02	9.33	8.11	8.45	15.4	15.3
1973-74	6.85	8.19	5.77	7.43	15.1	13.7
1972-73	7.19	7.07	6.55	6.16	12.3	14.1
1967-72	7.39	7.26	7.21	6.79	8.95	11.42
1962-67	5.16	4.34	4.73	4.09	9.37	6.96
1957-62	4.94	4.09	4.73	3.73	7.85	8.66
1952-57	5.06	4.79	4.74	4.47	9.59	9.87

SOURCE: See Table 2.5.
NOTE: See Table 2.5.

TABLE 2.7 Growth per 1% Increase in CPI

	Total Compensation		Wages and Salaries		Supplements	
Year	State and Local	Total	State and Local	Total	State and Local	Total
1977-78	0.94	1.02	0.77	0.94	1.83	1.45
1976-77	1.17	1.18	0.90	1.05	2.76	2.00
1975-76	1.42	1.40	1.17	1.23	3.09	2.50
1974-75	0.99	1.02	0.89	0.93	1.69	1.67
1973-74	0.62	0.75	0.53	0.68	1.37	1.25
1972-73	1.15	1.14	10.5	0.98	1.97	2.26
1967-72	1.60	1.57	1.56	1.47	1.94	2.48
1962-67	2.59	2.18	2.38	2.06	4.71	3.50
1957-67	3.41	2.82	3.26	2.57	5.07	5.97
1952-57	4.29	4.06	4.02	3.79	8.13	8.36

SOURCE: See Table 2.5.
NOTE: See Table 2.5.

It appears, then, that state and local government adjusted to the fiscal pressures of the mid-1970s by slowing the increase in employment, limiting wage increases, and actually decreasing wages in real terms. In fact, over much of the period 1972-1977, expenditures for personnel fell from 65% of state and local budgets (excluding capital expenditures) in 1972 to 54% in 1977. Thus, state and local employees are receiving a smaller and smaller share of state and local budgets.

This brief discussion has avoided many of the more complex issues in compensation of public employees, such as differences by level of government—federal employees have received consistently higher pay levels than their state and local counterparts. Differences in pay by function were also not included, in part because they are quite consistent with the overall trends.

Another topic in compensation of potential importance for urban governments is the role of unions in the public sector and their impact on wages. Since the demand for public employees is probably more inelastic than the demand for private sector employees, the potential effect of unions on wage increases (with minimal reductions in employment) is considerable. There are, however, more restrictions placed on unions in the public sector where strikes are generally illegal. Despite restrictions, the growth in union membership of state and local employees is phenomenal. Between 1960 and 1970, their membership more than doubled—from 361,000 to 796,000.

The effect of unions on wage increases in the public sector is not clear-cut. The relatively recent evidence contributed by Shapiro (1978) and Hamermesh (1975) suggests that unions increase wages of members relative to nonmembers to a degree roughly equal to unions in the private sector. That still implies a significant increase over nonunionized employees. Thus, if unionization continues to increase at previous rates there may be upward pressure on wages.[7]

SUMMARY

An analytic approach to explaning the determinants of public and service sector behavior in times of growth and decline requires a dynamic theoretical and empirical model. The theoretical models constructed by Baumol and Wagner were not particularly helpful in explaining the behavior of public employment during the 1970s, nor are they likely to be helpful in predicting that behavior for the 1980s. These models were constructed to explain sectoral behavior in periods of growth and not decline.

The "positive" analyses of the public choice school have also failed to explain the patterns of the 1970s. The models of public employee political power and budget maximization of bureaucrats were intended to explain the phenomenal expansion of the public, relative to the private, sector observed during most of the 20th century in the United States and were not equipped to deal with the effects of a nongrowing economy.

What in fact may be needed in the future is a new approach based on a theory of disequilibrium.[8] In the absence of such a model, significant insight can be gained from examining time series data on employment and compensation over the recent past.

The most important conclusions that emerge from the data presented here are:

(1) The pre-1972 expansion in public and service employment and compensation was impressive. The 1960s and early 1970s showed significant wage catch-up for state and local public employees relative to the private sector.

(2) The rate of growth in local government, full-time equivalent employment began to decline in the early 1970s. Nevertheless, even at lower growth rates the state and local sector was mildly employment stabilizing for the economy during the 1974-1975 recession.

(3) Employment in nongovernmental services appears to be procyclical, rising slowly during the recession and recovering rapidly after the recession. Full-time equivalent employment in the state and local sector, on the other hand, was not destabilizing over the period 1948-1978.

(4) After 1972, the state and local sector began to lag behind the private sector in improvement in compensation. This was true, with occasional yearly exceptions, for both wages and salaries and total compensation. Supplements, however, improved more rapidly than private sector supplements after 1972.

The evidence presented here indicates that the economics of decline (or disequilibrium) are not as clear-cut as the patterns of growth. There are also marked differences in the behavior of services and government during periods of decline, the former relatively procyclical and the latter less so. The trends in governmental employment, however, are more complex—there remains the issue of the significant slowdown in growth in local government after the recent recession.

Urban economies will be dominated by a more service-oriented employment base in the near future. These parts of the employment base were relatively volatile over the last 30 years. But the local government employment sector has recently exhibited employment patterns suggesting a shift away from public employment within the broader service and public sector aggregate. If that recent trend continues, local employees will become less dominant in the broader service sector. In compensation, state and local employees receive compensation that has remained quite close to the average for employment as a whole, with slightly slower rates of increase than the average (except for 1975-1976) since 1973.

This apparent slowdown in local government employment growth and state and local compensation growth could be a short-term phenomenon, or it could reflect the net effect of regional shifts in the economy. But there are certain to be painful adjustments, especially in urban areas already experiencing economic decline.

Other potential problems for urban economies in the 1980s seem likely. The growth in state and local supplements bodes ill. Local governments could be reacting to employee pressures by increasing pensions, therefore postponing "effective" wage increases until after employees retire. Given the pervasiveness of pension underfunding by local governments, the 1980s will present increasing fiscal pressure from these "uncontrollable" espenses. The increase in unionization in the public sector could also increase pressure on urban governments to increase compensation. As a balance, however, there has been a growing tax consciousness on the part of local taxpayers, which will likely continue to be a dampening influence on growth in local governments.

NOTES

1. This argument, as formulated by Shibata (1973), states that public employees have an incentive (and some monopsony power) to overstate their preferences for public goods, since that will increase their employment and, possibly, compensation. See Courant et al. (1979) for evidence on the limitations of that monopsony power.
2. If we reverse the arguments, Wagner's model implies that government would contract more than the goods-producing sector, which is a possible outcome in a declining economy for the Baumol model also.
3. Services refer to the industrial classification. This does not include all manufacturing. For example, trade, finance, insurance and real estate, transportation, communication, and public utilities are not included in this classification.
4. For a discussion of static stability, see Booth (1975). For more evidence on the cyclical stability of other parts of the new economic base, see McHugh et al. (forthcoming).
5. See Harrison and Hill (1978) for a good discussion of the changing structure of jobs in older and younger cities in the United States.
6. Fogel and Lewin (1974) maintain that the prevailing wage rate rule as implemented by the federal government is upwardly biased. Taking a more comprehensive view, Smith (1977) analyzes the relative wage rates of private and all levels of public employees. She finds that local public employees are *not* "overpaid" relative to private employees with the same skills.
7. Another speculation is that unions may exchange increases in supplements for increases in wages during periods of economic decline.
8. Frey (1970) hypothesized that the public sector will exhibit a cyclical pattern due to the separation of benefits and costs in the provision of public goods. In his model, the 1970s (and possibly the 1980s) would be a cost-conscious period characterized by contraction.

REFERENCES

ACIR (1978) Trends in Fiscal Federalism, Washington, DC: ACIR.
BAHL, R. W., D. GREYTAK, A. K. CAMPBELL, and M. J. WASYLENKO (1977) "Intergovernmental and functional aspects of public employment trends in the Unites States." Public Admin. Rev. 32: 818-832. November/December 1977, 818-32.
BAUMOL, W. J. (1967) "Macroeconomics of unbalanced growth; the anatomy of urban crisis." Amer. Econ. Rev. 57: 415-26.
BOOTH, D. E. (1975) "The differential impact of manufacturing activity on local government expenditures and revenues." National Tax J. (March 1975): 33-34.
BORCHERDING, T. E. (1977) "The Sources of growth of public expenditures in the United States, 1902-1970," in T. E. Borcherding (ed.) Budgets and Bureaucrats; The Sources of Government Growth. Chapel Hill, NC: Duke University Press.
---, W. C. BUSH, and R. M. SPANN (1977) "The effects on public spending of the divisibility of public outputs in consumption, bureaucratic power, and the size of the tax-sharing group," in T. E. Borcherding (ed.) Budgets and Bureaucrats: The Sources of Government Growth. Chapel Hill, NC: Duke University Press.
BRADFORD, D. F., R. A. MALT, and W. E. OATES (1969) "The rising cost of local public services: some evidence and reflections." National Tax J. 22: 185-202.

BUSH, W. C., and A. T. DENZAU (1977) "The voting behavior of bureaucrats and public sector growth," in T. E. Borcherding (ed.), Budgets and Bureaucrats: The Sources of Government Growth. Chapel Hill, NC: Duke University Press.
COURANT, P. N., E. M. GRAMLICH, and D. L. RUBINFELD (1979) "Public employee market power and the level of government spending." Amer. Econ. Rev. 69: 806-817.
EHRENBERG, R. (1973) "The demand for state and local government employees." Amer. Econ. Rev. 63: 366-79.
FOGEL, W., and D. LEWIN (1974) "Wage determination in the public sector." Industrial and Labor Relations Rev. 27: 410-31.
FREY, B. (1974) "A dynamic theory of public goods." Finanzarchiv 188-193.
HANSEN, A. H., and H. S. PERLOFF (1944) State and Local Finance in the National Economy. New York: W.W. Norton.
HAMERMESH, D. (1975) "The effect of government ownership on union wages," pp. 227-255 in D. Hamermesh (ed.) Labor in the Public and Nonprofit Sectors. Princeton, NJ: Princeton University Press.
HARRISON, B. and E. HILL (1978) "The changing structure of jobs in older and younger cities." Joint Center for Urban Studies of the Massachusetts Institute of Technology and Harvard University, Working Paper 58.
McHUGH, R., M. GELLERSON, and S. GROSSKOPF (forthcoming) "Cyclical stability of the new urban economic base," Northeast Regional Science Review.
NISKANEN, W. A. (1971) Bureaucracy and Representative Government. Chicago: Aldine-Atherton.
RAFUSE, R. W., Jr., (1965) "Cyclical behavior of state-local finances," in R. Musgrave (ed.) Essays in Fiscal Federalism. Washington, DC: Brookings Institution.
REDER, M. (1978) "The theory of employment and wages in the public sector," pp. 1-48 in D. Hamermesh (ed.) Labor in the Public and Nonprofit Sectors. Princeton, NJ: Princeton University Press.
ROSS, J. (1978) Countercyclical Aid and Economic Stabilization. Washington, DC: ACIR.
SHAPIRO, D. (1978) "Relative wage effects of unions in the public and private sectors." Industrial and Labor Relations Rev. 31: 193-204.
SHIBATA, H. (1973) "Public goods, increasing cost, and monopsony: comment." J. of Pol. Economy 81: 223-30.
--- (1971) "A Bargaining Model of the Pure Theory of Public Expenditures," Journal of Political Economy 79, January/February 1971, 1-29.
SMITH, S. (1977) Equal Pay in the Public Sector: Fact or Fantasy? Princeton, NJ: Princeton University Press.
U.S., Bureau of the Census (1979a) 1977 Census of Governments, Historical Statistics on Governmental Finances and Employment, vol. 6 no. 4. Washington, DC: Government Printing Office.
--- (1979b) Public Employment in 1978. Washington, DC: Government Printing Office.
U.S., Department of Commerce, Bureau of Economic Analysis (1976) Survey of Current Business, vol. 59 no. 4. Washington, DC: Government Printing Office.
WAGNER, A. (1877) Finanzwissenschaft, part 1. Leipzig: C. F. Winter.

3

The Prospects for Urban Revival

JAMES W. FOSSETT
University of Michigan
RICHARD P. NATHAN
Princeton University

☐ THE LAST FEW YEARS have seen a major shift in the rhetoric surrounding public discussion of the future of older American cities. By contrast with earlier prevailing wisdom, which had defined the older city as an obsolete economic life form, much reporting on urban conditions has been optimistic. Some observers have argued that decreases in family size and increases in energy and housing costs have made central cities more attractive places to live and point to increasing levels of housing rehabilitation in many cities as evidence that households are already moving in. Still other observers point to near boom levels of downtown construction, increased demand for office space at record rents, and the success of such commercial developments as Ghirardelli Square in San Francisco and Water Tower Plaza in Chicago as evidence that the economies of many older cities are beginning to revive (Allman, 1978; von Eckardt, 1979; U.S. News and World Report, 1980).

AUTHORS' NOTE: *We wish to acknowledge exemplary research assistance from Jerry P. Cawley of Michigan and Arthur Maurice of Princeton, help in obtaining information from Tim Jones of the Census Bureau; and useful comments on an earlier draft from Thomas J. Anton, Robert F. Cook, and Michael H. Schill. The research reported in this chapter was supported by the Department of Housing and Urban Development under Grant H2899RG to the Brookings Institution. The views and opinions in this chapter are those of the authors and should not be attributed to HUD, Brookings, Michigan, Princeton, or any other institution.*

In spite of these optimistic reports and the physical evidence of construction in Baltimore's Inner Harbor and Chicago's North Shore, a doubling of median office rents in Manhattan in the last five years, and impressive amounts of housing rehabilitation in such areas as Adams-Morgan in Washington and Queen Village in Philadelphia, it is far from clear that many older cities have turned the corner or even bottomed out. Some observers have argued that increased energy costs will not make urban locations appreciably more attractive either to residents or businesses, and may even result in increased movement out of cities (Small, 1980). Several careful studies in a number of cities from a variety of sources have failed to turn up much evidence of a "back to the city" or even a "remain in the city" movement of any appreciable size (Nelson, 1978; Long, 1980; Long and Dahmann, 1980; Schill, forthcoming). Further, it seems possible that the oft-noted reduction in urban household size represents a postponement of marriage and child bearing rather than a permanent shift in family patterns. The 1980s may see an acceleration of migration rates as the children of the baby boom years move through their own child bearing period (Barabba, 1980). Others have argued that current booms in downtown construction and real estate, while positive in themselves, are only local improvements that at best have little effect on the rest of the city and at worst have created problems elsewhere by driving out smaller businesses unable to pay increased rents (Sternlieb and Hughes, 1979; Vitullo-Martin, 1979). Advocates of downtown development respond to these claims by arguing that large anchor developments are necessary for other sectors to revive, and that this process is simply not far enough along for these broader benefits to have appeared.

While there has been considerable debate about the current state of American cities, there is relatively little concrete information about what has changed and by how much. As we have argued elsewhere, much of the debate over the recent past and future of American cities has been conducted using archetypes rather than evidence. Examples of rehabilitated neighborhoods in Adams-Morgan and Capitol Hill in Washington are countered with examples of residential displacement and further deterioration in Anacostia; and the West Side in Chicago is set against success stories in Lincoln Park and the North Shore. There is not much hard evidence, however, on how many housing units have been rehabilitated in Adams-Morgan, or in Washington as a whole; how many people have been displaced from Capitol Hill into Anacostia, or what effect Water Tower Plaza has had on Chicago's retail sales. Evidence on the net effect, either positive or negative, of the processes under way in either of these or other cities is a scarce commodity, and it is an open question whether such evidence is even collected.

This chapter addresses three questions about the nature and direction of urban change and the adequacy of available information on trends and conditions in cities. In the first section, we review the most current available evidence on the recent demographic and economic history of older American cities. Our conclusions here are pessimistic—there is little evidence that residential or economic revitalization has had any appreciable effect on any appreciable number of older cities. The second section assesses the adequacy, reliability, and timeliness of available data on economic and demographic trends in cities. Our conclusion is also pessimistic—we lack the statistical capability to say a great deal about recent changes in cities, and it is unlikely that we will gain that capability in the future. Finally, we advance some ideas about the questions future research on urban conditions should deal with and the most useful types of information for answering them.

TRENDS IN URBAN CONDITIONS

In an earlier paper, we presented evidence on changes in the concentration of urban problem conditions and economic growth over the 1960s for approximately sixty of the nation's largest cities (Nathan and Fossett, 1979). In this chapter, we examine more recent evidence on social and economic trends in this same set of cities in the 1970s to see whether the relative severity of urban problems or the level of relative economic growth have changed.

Much of our earlier work measured the level of urban hardship by the combination of three variables—city age (as indicated by the amount of older housing), population loss, and the concentration of poverty. The first two columns of Table 3.1 combine these factors into "urban conditions indexes" for 1960

66 URBAN GOVERNMENT FINANCE

TABLE 3.1 Changes in Urban Conditions, 1960-1977

Quintiles: 1960 UCI	City	Urban Conditions Index 1960	Urban Conditions Index 1970	Percentage Population Change (1960 base year) 1960 to 1970	Percentage Population Change 1970 to 1977	Percentage Population Change 1970 to 1975	Percentage Population Change 1975 to 1977	Per Capita Income 1959	Per Capita Income 1969	Per Capita Income 1974
		(1)	(2)	(3)	(4)	(5)	(6)	(7)	(8)	(9)
	ST LOUIS	207.6	232.6	-17.0	-13.9	-13.0	-1.0	1001	2726	4006
	BOSTON	201.0	193.2	-8.1	-3.2	-.6	-2.6	1919	3093	4157
	NEWARK	196.3	207.0	-5.6	-14.4	-10.6	-3.8	1792	2492	3348
	BUFFALO	190.9	244.1	-13.1	-13.6	-10.4	-3.2	1911	2877	4207
	PITTSBURGH	178.4	200.3	-13.9	-12.9	-10.2	-2.7	1943	3071	4436
I	CLEVELAND	177.9	215.7	-14.3	-16.2	-12.8	-3.4	1856	2821	3925
	PHILADELPHIA	166.2	168.5	-2.7	-8.5	-6.6	-1.9	1875	3017	4330
	ROCHESTER	166.2	187.9	-7.0	-12.5	-9.1	-3.4	2068	3239	4335
	DETROIT	154.0	151.9	-9.5	-13.3	-10.6	-2.7	2006	3200	4463
	CINCINNATI	145.2	149.6	-9.9	-9.8	-8.0	-1.8	2042	3132	4517
	MINNEAPOLIS	144.5	154.7	-10.0	-15.3	-11.7	-3.7	2247	3483	5161
	BALTIMORE	144.4	153.9	-3.5	-10.8	-5.8	-5.0	1867	2876	4330
	BIRMINGHAM	140.9	134.2	-11.8	-5.4	-7.2	1.8	1567	2368	4023
	TOLEDO	140.2	102.9	20.8	-8.0	-5.1	-2.8	2013	3252	4571
	LOUISVILLE	140.0	138.3	-7.5	-9.9	-6.5	-3.3	1761	2958	4302
	CHICAGO	138.6	146.9	-5.3	-8.4	-7.4	-1.0	2294	3402	4689
II	NEW ORLEANS	130.6	137.4	-5.4	-5.1	-5.4	.2	1739	2705	3268
	ST PAUL	127.6	132.2	-1.1	-14.0	-9.7	-4.3	2173	3397	4931
	NEW YORK CITY	127.6	117.8	1.5	-7.7	-5.3	-2.4	2306	3698	4939
	KANSAS CITY	124.5	102.8	6.7	-10.2	-7.3	-2.9	2174	3329	4736
	OAKLAND	120.7	106.6	-1.6	-8.0	-8.4	.5	2362	3616	5034
	AKRON	118.5	130.6	-5.1	-10.7	-8.2	-2.6	2122	3274	4614
	PORTLAND, OR	117.5	112.4	2.5	.5	-6.7	7.3	2281	3533	5192
	SAN FRANCISCO	115.0	116.4	-3.3	-8.2	-6.9	-1.3	2710	4232	5990
	MILWAUKEE	105.6	127.1	-3.3	-8.6	-6.9	-1.7	2104	3184	4680
III	OMAHA	99.0	87.1	15.2	6.1	8.0	-1.9	2135	3269	4887
	COLUMBUS	98.3	80.2	14.4	-1.5	-.8	-.7	1884	3025	4333
	MIAMI	83.5	65.7	14.9	4.0	10.3	-6.3	1834	2821	4416
	MEMPHIS	83.3	46.7	25.4	9.0	7.6	1.4	1650	2793	4303
	SAN ANTONIO	80.0	68.1	11.3	23.7	20.2	3.4	1426	2421	3601
	SEATTLE	79.1	87.7	-4.7	-7.5	-7.9	.3	2522	4052	5000
	DENVER	78.9	79.2	4.2	-8.0	-6.1	-1.9	2272	3534	5585
	OKLAHOMA CITY	71.4	56.6	13.1	1.6	-.3	1.8	1980	3236	4731
	ATLANTA	70.7	67.0	2.0	-16.5	-12.5	-4.0	1932	3156	4527
	NORFOLK	65.1	77.0	1.0	-8.5	-7.0	-1.6	1719	2792	4233
	FORT WORTH	64.5	53.2	10.4	-7.1	-9.9	2.7	1945	3236	4527
IV	BATON ROUGE	61.1	47.5	8.9	93.3	84.3	9.0	1850	2846	4187
	LOS ANGELES	57.9	51.1	13.6	-2.2	-3.6	1.4	2624	3951	5277
	AUSTIN	57.3	29.9	35.0	38.3	26.4	11.8	1683	2998	4379
	SACRAMENTO	56.7	43.9	32.7	5.3	3.4	1.9	2468	3383	4765
	TULSA	53.6	41.6	26.8	1.0	-.0	1.0	2293	3492	6473
	WICHITA	51.9	58.3	8.6	-3.4	-4.6	1.2	2077	3259	4951
	CHARLOTTE	51.0	35.1	19.7	27.0	19.9	7.1	1970	3272	4926
	LONG BEACH	50.4	54.5	4.2	-6.4	-6.7	.3	2458	3960	5652
	EL PASO	44.2	58.0	16.5	26.4	22.9	3.5	1576	2390	3479
	TAMPA	43.6	72.5	1.0	-4.7	.9	-5.6	1720	2779	4362
	HOUSTON	40.2	27.7	31.4	29.2	10.1	19.1	2062	3383	5110
V	DALLAS	38.9	28.1	24.2	.0	-4.6	4.7	2217	3697	5285
	SAN DIEGO	33.8	36.1	21.5	18.0	13.5	4.5	2301	3517	5016
	SAN JOSE	27.7	13.3	118.7	67.0	53.5	13.6	2204	3394	4972
	ALBUQUERQUE	18.6	23.9	21.2	23.6	17.7	5.9	2107	3091	4544
	PHOENIX	9.8	18.5	32.4	23.4	18.9	4.5	2013	3253	4719
	TUCSON	8.6	26.2	23.5	18.0	15.7	2.2	1879	2880	4385
	AVERAGE FOR CITIES	100.00	100.0	7.9	1.9	1.3	0.6	2033	3185	4635

The 1970 index is computed from the following formula:

$$\frac{\text{Mean per capita income 1970}}{\text{Per capita income 1970}} \times \frac{\text{Percent pre-1940 housing in 1970}}{\text{Mean percent pre-1940 housing in 1970}}$$

$$\frac{100 + \text{rate of population change 1960-70}}{100 + \text{median rate of population change 1960-70}}$$

The 1960 index is computed from the same formula as the 1970 index using 1960 data. Population change was measured from 1950 to 1960.

SOURCES: See p. 67.

and 1970 for 53 of the 57 largest cities in the country. Per capita income has been substituted for percentage of population in poverty in both years. Both indexes have been standardized

SOURCES: 1950, 1960 population and 1959 income: *County and City Data Book: 1962.* 1960 housing: *1960 Census of Housing, Volume 1: States and Small Areas.* 1970 population and 1969 income: *County and City Data Book: 1972.* 1970 housing: *1970 Census of Housing, Volume 1: States and Small Areas.* The preceding are from the U.S. Bureau of the Census (Washington, D.C., Government Printing Office). 1975 population and income: Data elements tape for Entitlement Period 10, furnished by the Office of Revenue Sharing, Department of the Treasury. 1977 population: *General Revenue Sharing Initial State and Local Data Elements Entitlement Period 11,* Office of Revenue Sharing, Department of the Treasury.

to a mean of 100.[1] Cities with index scores above 100 are "worse off"—more distressed—than cities with lower scores.

Two general points suggest themselves. First, there is a substantial correlation (.96) between indexes in these two years; cities with problems in 1960, as measured by this index, continued to have them through 1970. All of the 25 cities with index scores above the mean in 1960 had scores above the mean in 1970, and all the cities with index scores below the 1960 mean had similar scores in 1970.

More importantly, there was an increasing disparity in the concentration of urban problems over the 1960s. Of the 25 cities with index scores above the mean in 1960, 16 had become even more distressed as measured by this index in 1970. By contrast, of the 28 cities with scores below 100 in 1960, 17 improved their relative position by 1970. The most distressed cities in 1960, according to this index, lost an average of 10% of their population between 1960 and 1970, while the populations of the least distressed cities grew, on average, by almost 30% over this same period. Similarly, per capita income in the most distressed cities grew at a slower rate during the 1960s than it did in more prosperous cities; the income gap between more and less prosperous cities increased. Quite literally, over this period, the rich got richer and the poor got poorer.

Although the verdict still is not in, the available evidence suggests that this gap between rich and poor cities has increased substantially during the 1970s, particularly during the first half of the decade. Columns 3 through 6 of Table 3.1 present information on changes in population between 1960 and 1977 in these cities. To make comparison easier, all these figures are expressed as changes from the 1960 population.

Table 3.2 presents population change between 1960 and 1977 as a percentage of 1960 population for each quintile of the 1960 urban conditions index (UCI), arranged from most to

TABLE 3.2 Average Population Change, 1966-1977, as Percentage of 1960 Population, by 1960 UCI Quintile

Quintiles, 1960 UCI	Average Percentage Population Change (1960 base)			
	1960-1970	1970-1977	1970-1975	1975-1977
I[a]	-10.1	-12.2	-9.4	-2.7
II[b]	-1.1	-8.9	-6.9	-2.0
III[c]	7.7	0.9	1.1	-0.1
IV[d]	15.2	10.2	7.6	2.5
V[e]	28.6	20.2	14.7	5.4
All cities	7.9	1.9	1.3	0.6

SOURCE: See Table 3.1.

[a] First quintile (most distressed): St. Louis, Boston, Newark, Buffalo, Pittsburgh, Cleveland, Philadelphia, Rochester, Detroit, Cincinnati, Minneapolis.
[b] Second quintile: Baltimore, Birmingham, Toledo, Louisville, Chicago, New Orleans, St. Paul, New York City, Kansas City, Oakland, Akron.
[c] Third quintile: Portland, San Francisco, Milwaukee, Omaha, Columbus, Miami, Memphis, San Antonio, Seattle, Denver.
[d] Fourth quintile: Oklahoma City, Atlanta, Norfolk, Fort Worth, Baton Rouge, Los Angeles, Austin, Sacramento, Tulsa, Wichita.
[e] Fifth quintile (least distressed): Charlotte, Long Beach, El Paso, Tampa, Houston, Dallas, San Diego, San Jose, Albuquerque, Phoenix, Tucson.

least distressed. These figures suggest that cities classified as more distressed lost population at a faster rate between 1970 and 1977 than during the 1960s, while more prosperous cities grew at roughly the same rate as they had earlier. Cities falling in the top (most distressed) quintile of the urban conditions index in 1960 lost an average of 10.1% of their 1960 population between 1960 and 1970. Between 1970 and 1977, they lost, on the average, an additional 12.2% of their 1960 populations. Cities in the bottom (least distressed) quintile of the 1960 index, whose populations had increased an average of 28.6% between 1960 and 1970, experienced a further average increase of over 20% of their 1960 populations between 1970 and 1977. In short, at least in the aggregate, the rate of population loss in more distressed cities over the 1970s got worse, not better.

This disparity in growth narrowed slightly between 1975 and 1977, the most recent period for which data are available. Cities classified as more distressed in 1960 lost an average of 1.37% of their 1960 populations in each of the two years from 1975 to 1977, a slight decline from the period 1970-1975, when they lost an average of 1.89% of 1960 population. Cities in the second quintile of the urban conditions index in 1960 lost an annual average of just under 1% of 1960 population between 1975 and 1977, an improvement from 1970-1975, when the annual population loss averaged 1.4%.

While the more distressed cities were losing population at a slower rate, the more prosperous cities were gaining it at a slower rate during this period; they showed a drop in the rate of growth of about one-quarter of a percentage of 1960 population a year. In short, the disparity in population growth between more and less prosperous cities declined between 1975 and 1977 by almost 20% from an annual average of almost 5% of 1960 population between 1970 and 1975 to slightly over 4%.

Many inferences have been drawn from these figures. Some observers have argued that the reduction in the rate of population decline in distressed cities is itself a favorable sign, signalling a reduction in migration rates stemming from declines in marriages and childbearing. Others have argued that the critical variable in assessing trends in urban demography is no longer change in the number of people, but rather change in the number of households, which is reported to be increasing in many cities. For both arguments, the conclusion is optimistic: In spite of the loss of population, the number of households seeking accommodations has remained level or increased. As a result, both city housing markets and tax bases may be reviving.

In our view, this scenario may be overly optimistic. First, reports of the demographic rebirth of older cities appear to be premature. The rate of population loss in the more distressed cities does appear to have declined over the mid-1970s, but through 1977 these cities were still losing population at a faster rate than during the 1960s. Of the 22 most distressed cities in 1960, 17 lost population between 1975 and 1977 at a faster rate than they did between 1960 and 1970. This deterioration was particularly pronounced in several cities labelled by many observers, including us, as candidates for revival. Boston, for example, lost an annual average of 0.81% of its 1960 population over the 1960s, between 1975 and 1977, it lost an average of 1.3% per year. This contrast was even more marked in Baltimore, whose annual rate of decline increased by over two percentage points, from 0.35% per year in the 1960s to 2.5% between 1975 and 1977. The movement of people out of older cities may have slowed, but it has not stopped.

Second, the increase in the number of households—which some say is the salvation of older cities—does not appear to have been large enough—at least through 1977—to maintain housing demands in more distressed cities. Nationally, the number of households in *all* 243 central cities increased by 1.8 million (6.3%) between 1970 and 1977 (Sternlieb and Hughes, 1979: 630). We have no information on changes in this figure for individual cities and will not have until reports from the 1980 census are available. It seems likely, however, that most of this increase occurred in more prosperous cities, where populations increased by 20% over the 1970s, and in smaller central cities not included in this study. Many smaller cities experienced substantial population growth over the 1970s, and it seems likely that many new households were formed in these places. There is no reason to suspect that the forces producing smaller households are limited to more distressed cities, making it possible, if not likely, that household growth was distributed roughly in the same geographic fashion as population growth. The growth in households in more distressed cities has probably been relatively small and may not have taken place at all in many cities.

A final reason for skepticism about prospects for residential revival in distressed cities is that if there has been an increase in the number of households in older cities—and it is worth reemphasizing that we have no information one way or the other on this question—the limited aggregate data available suggest the source of this increase has *not* typically been professional couples buying condominiums in Lincoln Park in Chicago or GS-11s moving into group houses in Adams-Morgan in Washington. The household types that have been growing the fastest in urban areas are those headed by females or those composed of unrelated individuals. These two types of household are generally poorer than more traditional family households. In addition, the absolute numbers and the concentration of individuals with incomes below the poverty level increased in central cities between 1970 and 1977. While we have no evidence for individual cities on this question either, it seems at least a plausible inference that many, if not most, of whatever new households there may be in older cities are relatively poor (Sternlieb and Hughes, 1979: 630-631; Shalala, 1979: table 11). While higher income households may in fact be settling in some more distressed cities, there is no evidence as yet to suggest that this movement has been large enough to produce any net improvement.

TRENDS IN URBAN ECONOMIC GROWTH, 1958-1977

The prospects for economic revitalization in older, more distressed cities appear no more favorable than those for major residential improvement. Although the disparity in economic growth rates between richer and poorer cities narrowed slightly in the middle 1970s, this narrowing appears to be due to a recovery from an extremely severe economic slump in the late 1960s and early 1970s, rather than to any nascent revival. Jobs and economic activity continue to leave more distressed cities at a substantial rate albeit slightly slower than in earlier periods.

Our earlier work measured economic growth in cities by a "composite economic indicator." This index measures growth in local manufacturing, wholesale and retail trade, and services, weighted by employment in each sector, relative to the average change for all cities (Table 3.3). As with earlier indexes, this measure has been standardized to a mean of 100, so that cities with slow growth rates have higher scores.

While there are some differences between this measure of economic growth and the measures reported earlier, the overall pattern is similar. The composite economic indicator is correlated with the 1960 urban conditions index at .77 and with the 1970 index at .80. Not surprisingly, cities with slow rates of economic growth also tended to have lower per capita incomes, larger concentrations of poor, and higher rates of population loss.

These figures also reinforce the earlier argument that the disparity between rich and poor cities increased during the 1960s. Cities with urban conditions index scores in the top (most distressed) quintile of the 1960 index had an average growth rate score between 1958 and 1972 of over 200, suggesting a growth rate of less than half the average for all cities in this period. Cities with 1960 index scores in the second quintile had an average growth score of just over 100, while cities with 1960 index scores in the bottom three quintiles had growth rates above the average.

This disparity in economic growth does not appear to have occurred uniformly, but to have been concentrated during the latter part of this 14-year period. Columns 2 through 4 in Table

72 URBAN GOVERNMENT FINANCE

TABLE 3.3 Economic Growth Measures, 1958-1977

Quintiles, 1960 UCI	City	Composite Economic Growth Indicators					Percentage Change		Selected
		1958 to 1972	1958 to 1963	1963 to 1967	1967 to 1972	Manufacturing Value Added 1972-1976	Retail Sales 1972-1977	Wholesale Sales 1972-1977	Service Receipts 1972-1977
		(1)	(2)	(3)	(4)	(5)	(6)	(7)	(8)
I	ST LOUIS	245.9	140.9	121.2	288.6	37.8	17.5	73.0	12.8
	BOSTON	142.2	130.3	117.2	128.7	15.8	12.5	19.3	37.5
	NEWARK	431.5	150.2	202.9	288.1	18.9	-26.9	34.8	21.9
	BUFFALO	209.1	247.8	116.2	150.4	36.1	11.4	6.0	23.6
	PITTSBURGH	296.8	189.8	126.4	272.0	25.0	33.5	79.6	41.8
	CLEVELAND	192.7	122.5	153.3	158.7	28.8	17.4	16.8	14.6
	PHILADELPHIA	179.1	152.1	134.2	136.7	10.4	21.0	19.4	26.8
	ROCHESTER	75.4	68.4	61.0	136.2	32.5	5.7	-7.7	34.6
	DETROIT	129.2	84.9	133.4	150.2	24.8	9.0	12.5	-17.3
	CINCINNATI	150.4	130.0	85.9	192.1	63.1	33.8	55.2	35.1
	MINNEAPOLIS	149.8	86.0	179.4	150.9	61.0	25.1	36.8	30.3
II	BALTIMORE	126.6	123.9	119.4	114.4	33.3	15.3	19.8	20.9
	BIRMINGHAM	83.7	159.6	87.2	67.9	66.7	36.4	***	23.8
	TOLEDO	65.9	77.4	75.3	77.4	63.4	40.1	72.0	42.9
	LOUISVILLE	82.2	76.8	96.9	92.4	50.7	24.1	***	48.9
	CHICAGO	148.0	134.0	***	***	22.8	19.5	33.2	18.7
	NEW ORLEANS	90.9	107.9	89.0	94.0	42.7	47.9	24.3	58.9
	ST PAUL	79.4	83.1	80.6	94.2	45.6	33.6	65.5	34.3
	NEW YORK CITY	123.1	98.8	136.0	116.1	3.9	14.7	64.1	8.2
	KANSAS CITY	***	72.8	***	***	20.5	32.9	***	46.8
	OAKLAND	134.8	90.8	184.2	121.3	46.6	30.3	21.3	56.00
	AKRON	100.1	90.7	85.9	116.6	-7.1	48.0	28.3	43.1
III	PORTLAND, OR	***	***	***	***	55.3	38.9	***	60.6
	SAN FRANCISCO	103.6	104.1	133.7	90.9	33.3	42.9	48.7	59.4
	MILWAUKEE	144.0	121.5	132.4	119.2	34.0	37.5	50.3	32.4
	OMAHA	57.3	103.3	69.5	57.9	22.0	43.5	24.3	47.9
	COLUMBUS	57.1	57.6	107.0	59.1	33.8	47.6	106.4	45.3
	MIAMI	65.6	154.9	122.3	45.0	39.2	28.5	***	41.7
	MEMPHIS	41.1	65.7	73.8	44.6	27.4	49.9	***	40.5
	SAN ANTONIO	***	***	75.5	48.3	34.4	68.3	***	57.1
	SEATTLE	121.9	89.9	159.9	105.3	37.2	57.5	73.3	77.7
	DENVER	57.1	84.5	118.0	48.4	***	39.7	72.6	71.7
IV	OKLAHOMA CITY	***	***	***	***	24.5	61.7	***	69.5
	ATLANTA	51.3	57.1	88.0	61.8	37.2	3.1	***	30.8
	NORFOLK	60.1	65.9	124.6	58.0	52.0	29.9	87.1	32.7
	FORTWORTH	***	168.4	37.3	***	92.8	51.8	***	62.0
	BATON ROUGE	33.1	81.8	31.4	57.3	***	70.7	137.2	110.3
	LOS ANGELES	92.4	103.9	86.3	101.7	49.3	49.3	76.3	59.2
	AUSTIN	29.1	59.1	55.5	37.5	***	89.4	***	85.2
	SACRAMENTO	47.1	54.3	72.1	72.1	***	55.7	***	***
	TULSA	37.3	***	***	***	62.4	77.2	***	62.6
	WICHITA	56.1	128.5	***	***	67.6	66.9	97.8	67.6
V	CHARLOTTE	39.3	56.4	70.3	47.2	17.1	35.6	55.2	76.2
	LONG BEACH	44.4	157.6	32.3	64.4	-35.3	47.6	***	61.9
	EL PASO	44.9	101.0	61.0	44.6	44.1	71.5	***	78.8
	TAMPA	40.7	68.6	82.1	40.3	27.4	38.4	***	74.5
	HOUSTON	***	60.6	55.1	***	85.5	90.1	***	158.1
	DALLAS	49.4	64.4	72.3	60.3	52.2	56.1	***	69.2
	SAN DIEGO	67.4	105.3	123.4	49.8	56.2	70.3	89.8	78.9
	SAN JOSE	41.9	46.2	54.6	77.7	49.3	78.2	114.1	93.0
	ALBUQUERQUE	33.8	50.6	***	***	***	75.7	77.2	***
	PHOENIX	20.6	28.7	68.4	32.0	50.5	56.9	69.8	58.2
	TUCSON	26.5	41.1	107.8	29.4	***	57.1	40.0	45.7
	AVERAGE FOR CITIES	100.0	100.0	100.0	100.0	38.1	41.3	51.5	50.8

SOURCES: 1976 Value added: Bureau of Census *1975 Annual Survey of Manufacturing,* Table 2. Other 1977 figures; preliminary and final area reports from relevant 1977 Census of Business. All other data: *County and City Data Books,* 1963, 1967, 1972, 1977.

NOTE: The composite economic indicator (CEI) is calculated according to the following formula:

$$CEI = \overline{(PCVAM)*(PMFE)+(PCRS)*(PRSE)+(PCWS)*(PWSE)+(PCSVC)*(PSVCE)}$$ (Average growth, sample cities)

where:
- PCVAM = percentage change in value added by manufacturing, indicated period (e.g., 1958-1963)
- PCRS = percentage change in retail sales receipts, indicated period
- PCWS = percentage change in wholesale sales receipts, indicated period
- PCSVC = percentage change in selected service receipts, indicated period
- PMFE, PRSE, PWSE, PSVCE = percentage of total employment in manufacturing, retail sales, wholesale sales, and selected services employed in given sector, during the initial year of a given period

It was necessary to adjust the growth figures used to calculate the CEI to eliminate negative numbers. A detailed list of the adjustments made is available from the authors on request.

3.4 report composite economic indicator scores for individual cities for each of the three periods defined by the Census of Business. Table 3.4 combines these scores to report average economic growth rate scores for each quintile of the 1960

TABLE 3.4 Average Composite Economic Indicator Scores, 1958-1972, by 1960 UCI Quintile

Quintiles, 1960 UCI	Average Composite Economic Indicator 1958-1972	Avergae Composite Economic Indicator 1958-1963	Average Composite Economic Indicator 1963-1967	Average Composite Economic Indicator 1967-1972
I	200	137	130	187
II	103	101	106	99
III	81	98	110	69
IV	51	90	71	65
V	41	71	73	50

SOURCE: See Table 3.3.
NOTE: Quintile numbers correspond to those for Table 3.2.

urban conditions index for each four to five year period and over the entire fourteen year period.

These figures indicate a convergence in rates of economic growth between more and less prosperous cities between 1958 and 1967 and a major divergence between 1967 and 1972. The average economic indicator scores of cities in the top quintile of the urban conditions index declined between 1963 and 1967 from 137 to 130, indicating a growth rate between 1963 and 1967 closer to the national average in this period than during the earlier period. While more prosperous cities grew at even faster rates during this period, the disparity in economic growth was smaller than that between 1958 and 1963.

The reasons for this convergence are easy to suggest. The period between 1963 and 1967 fell in the middle of the most sustained economic expansion in recent history. The average annual rate of growth in GNP over this period was 4.9%, with three of the four years showing growth rates of over 5%; average annual growth in the period 1958-1963 was only 4.1% (Varaiya and Wiseman, 1978: 29). This relatively high rate of aggregate growth in the mid-1960s had a disproportionately favorable effect on the economies of more distressed cities.

By contrast, the disparity in economic growth rates between more and less distressed cities increased substantially between 1967 and 1972. The average economic growth scores of cities in the top quintile of the 1960 urban conditions index increased by almost 50%, indicating a sharp deterioration in their relative economic performance; the average growth scores of cities in the lower four quintiles declined, indicating an improvement in their relative economic position.

This pattern of economic improvement in more distressed cities between 1963 and 1967 and sharp deterioration in the next five years is even more pronounced in the employment trends in the four sectors contained in the growth index. Table 3.5 shows average total employment growth between 1958 and 1972 for the cities in each quintile of the 1960 urban conditions index. Although total employment in all cities grew at respectable rates between 1963 and 1967, the biggest relative improvement occurred in cities classified as most distressed in 1960. These employment gains were wiped out in the succeeding five-year period, when the more distressed cities lost an average of almost 15% of 1967 employment. Employment in

TABLE 3.5 Average Percentage Change in Total Employment in Major Sectors, 1958-1972, by 1960 UCI Quintile

Quintiles 1960 Urban Conditions Index	Average Percentage Change in Employment			
	1958-1963	1963-1967	1967-1972	1958-1972
I	-6.9	4.2	-15.0	-17.3
II	-3.7	8.4	-4.0	-1.1
III	0.5	6.9	10.1	16.3
IV	4.6	21.2	16.3	50.4
V	17.5	21.9	23.0	77.4
All Cities	2.4	12.0	4.4	23.5

SOURCE: See Table 3.3.
NOTE: Quintile numbers correspond to those for Table 3.2.

more prosperous cities continued to expand over this period, in many cases at rates even higher than over the preceding five years.

The reasons for this sharp divergence in economic performance between more and less prosperous cities are difficult to define precisely. Aggregate economic growth over this period was slow relative to the earlier period; annual GNP growth over this period was only about 3% a year, and there was a sharp, if relatively short, recession in 1969-1970. Further, some of the cities falling into the top quintile of the 1960 index—Newark, Detroit, and Cleveland—were the scenes of major riots. These factors may have accelerated employment movement already under way in these cities to produce this severe deterioration in economic growth and employment.

As of this writing, information is just beginning to be available that allows an assessment of more recent changes. Preliminary area reports from the 1977 Censuses of Business are now available for most cities for all sectors except manufacturing, and final reports are out for many cities. Although we cannot assess recent changes in detail until all these data are available, we can make some general conclusions about changes over the mid-to late 1970s. Columns 5-8 of Table 3.3 provide the information available as of March 1980 on changes since 1972 in retail and wholesale trade, service receipts, and value added by manufacturing. All data are either preliminary or final figures from the relevant 1977 census of business except for the value-added figure. Because no reports, either final or preliminary, have yet been issued on value added from the 1977 *Census of*

Manufacturing, we are forced to use 1976 data taken from the *Annual Survey of Manufacturing* for that year. Table 3.6 presents average changes in these measures for the quintiles of the 1960 urban conditions index. Comparable measures are also displayed for 1967 to 1972. These figures suggest that the economies of older cities did improve over the middle 1970s, but only relative to their earlier, depressed performance. The disparity in growth rates between more and less prosperous cities did narrow over this period, particularly in manufacturing and wholesale trade. The disparity in the rate of growth between the fastest and slowest growing sets of cities declined from over 65% in the period 1967-1972 to less than 25% in manufacturing. The gap became narrower in wholesale and retail trade. The disparity between faster and slower growing cities actually increased in services, but this increase probably reflects an expansion of coverage in the 1972 Census of Selected Services, when several new industries were added.[2]

Perfunctory examination suggests that the bulk of this "recovery" was purely statistical, reflecting an improvement from a severely depressed base in more distressed cities. As columns 1, 4, and 7 of Table 3.6 indicate, growth in manufacturing and wholesale and retail trade between 1967 and 1972 was extremely low in the most distressed cities—less than 1% in wholesale trade and less than 3% in manufacturing. Even without price adjustments, five of the eleven cities in the top quintile of the 1960 index produced less in manufacturing output in 1972 than in 1967. Wholesale sales were lower in 1972 than in 1967 in six cities in this group; 1972 retail sales were below 1967 levels in two cities. But growth rates in more prosperous cities between 1967 and 1972 were quite substantial. Manufacturing output grew by more than 70% in the most prosperous set of cities; wholesale sales expanded by almost 75%. The fact that growth rates in manufacturing and wholesale trade in these two sets of cities were closer together in the succeeding five-year period should not be taken as evidence of revitalization, but rather of a reduction in the rate of growth (at least in manufacturing) in more prosperous cities and of a modest recovery from an extremely depressed base in more distressed ones.

A more direct way of putting these figures in perspective is to examine growth in these four sectors between 1967, the time

TABLE 3.6 Average Percentage Change in Value Added, Sales, and Receipts, 1967-1972 and 1972-1976/1977, by 1960 UCI Quintile

Quintiles 1960 UCI	Value added by manufacturing			Wholesale sales			Retail sales			Selected service receipts		
	1967-1972 (1)	1972-1976 (2)	1967-1976 (3)	1967-1972 (4)	1972-1977 (5)	1967-1977 (6)	1967-1972 (7)	1972-1977 (8)	1967-1977 (9)	1967-1972 (10)	1972-1977 (11)	1967-1977 (12)
I	2.9	32.2	36.4	.5	31.4	32.4	5.6	14.5	21.3	58.6	25.6	100.8
II	19.4	35.4	62.5	22.2	32.0	57.0	22.4	31.2	60.9	75.6	36.6	144.0
III	43.6	35.2	88.6	41.1	62.6	110.9	38.2	45.4	102.6	101.0	53.4	208.0
IV	58.0	55.1	131.6	39.0	99.6	167.2	43.9	55.6	126.3	95.0	64.4	221.9
V	70.3	38.6	115.8	74.7	74.3	213.9	60.4	61.6	160.3	117.1	79.6	286.9
All Cities	38.0	38.1	80.8	34.7	51.5	98.0	33.8	41.3	93.3	88.9	50.8	190.0

SOURCE: See Table 3.3.

when these cities were the closest together, and 1977. These figures are displayed in columns 3, 6, 9, and 12 of Table 3.6. The disparity in economic growth between these groups of cities over this period was substantial. Value-added increased by just over one-third in the most distressed cities between 1967 and 1976; it more than doubled in the most prosperous ones. Retail and wholesale sales grew by smaller amounts than manufacturing in the most distressed cities and by larger amounts in the most prosperous places. The disparity in service receipts was smaller than in the other sectors, but this may have been the result of a shift in coverage rather than a difference in growth. A resolution of this question will have to await a full set of 1977 figures.

In short, there was some increase in the growth rate in the economies of more distressed cities—but only some. The disparity in growth rates between more and less prosperous cities declined slightly over the middle 1970s, but it was still substantial. Retail sales grew, on the average, four times as fast in the most prosperous cities as in the most distressed, and the service receipts in rich cities grew almost three times as fast as in the poor ones. Although these differences are smaller than those of the period 1967-1972, they can hardly be taken as evidence of commercial revitalization in the more distressed places.

Trends in employment in the four sectors of the economy show a similar pattern. We lack employment figures for all four sectors for all the cities, but enough information is available on enough cities to provide a reasonably clear picture of changes between 1972 and 1977. Available data are presented in Table 3.7; data from the preceding five years are listed for comparative purposes. These figures show a familiar pattern: The rate of decline slowed in the most distressed cities, at least in manufacturing and wholesale trade. The rate of growth also slowed in more prosperous cities in these same sectors. The rate of growth in employment in retail trade and services in more prosperous cities slowed as well, but the rate of job loss in retail trade and services accelerated or remained about the same in the cities in the top two quintiles of the 1960 index. As with the receipt numbers, we cannot isolate changes in service employment due to shifts in coverage. Other observers have, however, reported growth in service jobs between 1967 and 1972 in several dis-

TABLE 3.7 Average Change in Employment, 1967-1977, for Major Sectors, by 1960 UCI Quintile

Quintiles, 1960 UCI	Manufacturing		Wholesale trade		Retail trade		Selected services		Total employment four sectors	
	1967-1972 (1)	1972-1976 (2)	1967-1972 (3)	1972-1977 (4)	1967-1972 (5)	1972-1977 (6)	1967-1972 (7)	1972-1977 (8)	1967-1972 (9)	1972-1977 (10)
I	-22.6	-9.1	-19.9	-15.4	-11.3	-11.7	15.5	-4.2	-15.0	-9.4
II	-11.7	-9.8	-7.2	-16.1	-2.0	-2.8	22.8	6.0	-4.0	-7.3
III	2.4	-3.8	3.6	2.3	11.4	8.5	37.1	16.6	10.1	N.A.*
IV	13.6	0	9.3	N.A.*	13.3	15.9	36.8	23.9	16.3	N.A.*
V	20.3	2.1	19.9	18.6	25.3	21.3	53.6	N.A.*	23.0	N.A.*
All	-13.0	-4.7	0.3	-3.7	7.3	6.0	32.6	9.2	4.4	N.A.*

SOURCE: See Table 3.3.
*Less than five cases. 1976-77 Employment figures for all sectors are available for only seven of the thirty-one cities in bottom three quintiles.

tressed cities controlling for these shifts (Varaiya and Wiseman, 1978: 36).

In short, it seems reasonably clear that service employment in more distressed cities increased between 1967 and 1972 and declined in the succeeding five years. Local government employment, which also had been increasing in more distressed cities over the late 1960s, may also have declined during this period, even though millions of dollars for public service employment were pumped into cities during this period (Nathan et al., 1979: 11-13; Peterson, 1978).

Employment in the most distressed cities fell between 1972 and 1977 at an annual rate of about 2%; the decline in the preceding five years was 3% per year. This reduction in the rate of job loss, however, was limited to manufacturing and wholesale trade. The rate of job loss in retail trade in the most distressed cities was slightly higher during the middle 1970s than it had been during the preceding five years. Employment in services and government, which had been growing over the late 1960s and early 1970s, declined during this period.

The most optimistic construction that can be placed on these figures is that economic conditions improved only slightly in more distressed cities. During the mid-1970s, their rates of economic growth were slightly higher, and rates of job loss in some sectors lower, than earlier rates. But rates of growth were still lower and rates of job loss higher than in more prosperous cities.

This disparity was particularly pronounced in retail trade and services, sectors identified as growth industries in declining cities. Growth in retail trade in the most prosperous cities was almost four times higher than in the most distressed; growth in service receipts was three times higher. Employment in retail trade increased by more than 20% in the most prosperous cities; it fell by almost 12% in the most distressed ones. Service employment grew at levels ranging from 16% to 24% in more prosperous cities; it fell by over 4% in the most distressed ones. The gap in economic and employment growth between rich and poor cities may have narrowed slightly over the middle 1970s, but still remains substantial.

Reconciling these figures with the evidence of increased economic activity in many downtowns is difficult, but several explanations can be advanced. One is that these figures from

1977 are simply too old to capture a recent upturn. A second possible explanation is that the figures reported here are not accurate reflections of the state of affairs in most cities, because, in 1977, cities had still not fully recovered from the recession of 1974-1975. A third interpretation, which we favor, is that much current downtown activity is economic gentrification—the consolidation of existing activity into a few high-rent locations, accompanied by continued decline elsewhere.

Choosing between these interpretations at this point is largely a matter of faith rather than evidence. But it is worth noting that an economic revitalization of any consequence in older, more distressed cities since 1977 would require a major reversal of long-term trends of the sort that rarely happens.

This unfavorable picture persists even in the older cities—Baltimore, Boston, Chicago, and Philadelphia, among others—that we and other observers have identified as potential candidates for "breaking out" of decline. The rates of population loss in Baltimore and Boston actually accelerated between 1975 and 1977. The rates of income growth in Boston, Chicago, and Philadelphia between 1970 and 1975 were below average for the cities reported here. Perhaps more important, between 1970 and 1975, transfer income—social security, unemployment, and welfare payments—increased in all four of these cities or the counties overlying them at a level substantially above total income. This suggests that earned income, such as wages and salaries, may have grown at an even lower rate (Shalala, 1979) than the total income figures suggest.

Similarly, these cities' economic performance over the middle to late 1970s provides little support for hopes of impending revival. Table 3.8 displays changes in value added, sales, receipts, and employment between 1972 and 1977 for each of these cities; for cities in each of the top two, or most distressed, quintiles of the 1960 urban conditions index; and for the entire set of cities reported here. These figures provide little evidence that these cities are beginning to "break away."

Particular cities do show higher growth rates in particular sectors. But growth rates in these cities are generally below those for cities in the second quintile of the 1960 index and are, without exception, substantially below growth rates for the entire set of cities. Baltimore, Chicago, and Philadelphia had higher rates of growth in retail sales than cities in the top

TABLE 3.8. Economic Growth in Selected Cities, 1972-1976/1977

	Manufacturing Value Added 1972-1976	Percentage Change in:		Selected service receipts 1972-1977	Total employment four sectors 1972-1977
		Retail sales 1972-1977	Wholesale sales 1972-1977		
Baltimore	33.3	15.3	19.8	20.9	-15.8
Boston	15.8	12.5	19.3	37.5	-9.9
Chicago	22.8	19.5	33.2	18.7	-8.2
Philadelphia	10.4	21.0	19.4	26.8	-15.1
Cities in top (most distressed) quintile, 1960 Urban Conditions Index	32.2	14.5	31.4	25.6	-9.4
Cities in second quintile, 1960 Urban Conditions Index	35.4	31.2	32.0	36.6	-7.3
Total. 53 cities	38.2	41.3	51.5	50.8	N.A.*

SOURCE: See Table 3.3.
*Not reported due to limited number of cases. See note to Table 3.3.

quintile of the 1960 index. But the growth rate in Philadelphia, the highest of the three, was more than ten points below the average for cities in the second quintile and only slightly more than half the growth rate for all 53 cities. The same holds true for service receipts in Boston—they grew at a rate higher than the average for the most distressed cities but slower than the average for cities in the second quintile, and at only half the rate of all cities taken together.

Perhaps more important, overall rates of job loss in Boston, Baltimore, and Philadelphia were *above* the average for other distressed places. More detailed figures show that the rates of job loss in retail and service employment were slightly smaller for these cities than for other cities in the first quintile of the 1960 index. In every case the rates of employment decline in these four cities were higher than cities in the second quintile and considerably higher than the entire set of cities, reported here, in which employment in these two sectors increased substantially.

Chicago, for example, had the smallest rate of retail job loss of the four cities—7.6%. This rate is considerably lower than the average job loss of the most distressed cities (11.7%), but it is almost five points higher than the average for the cities in the second quintile of the 1960 index. Similarly, both Boston and Chicago experienced gains in service employment of 0.4%. Although this gain is higher than that experienced by other distressed cities, where average service employment fell by over 4%, it is over 5% lower than that experienced by the second-quintile cities, where service employment grew by an average of almost 6%; and considerably below the 10% average increase for all 53 cities.

In short, the cities alleged to be on the threshold of economic and demographic recovery show little evidence of having begun this process as of the late 1970s. Residential rehabilitation and new commercial developments have sprung up at various locations, but are still too small to boost the aggregate level of prosperity. The pattern that emerges from these figures is one of isolated improvements in a few "pockets of plenty," accompanied by continued decline elsewhere.

It might be argued that this scenario is too pessimistic. More recent figures, measured over a shorter period than five years, might show that things have improved in these places or at least quit declining. Several of the major developments alleged to serve as anchors for future improvement either were not open in 1977 or had not been operating long enough to show any appreciable impact on 1977 sales and receipts. There also have been reports of increasing levels of housing rehabilitation of late in such areas as Queen Village in Philadelphia and Lincoln Park in Chicago. If measured more recently, these cities might look better.

This alternate scenario is not impossible, but it is unlikely. The downward trend in employment and income in these cities is strong enough and the rate of economic growth is small enough that a sharp reversal of direction in less than three years is highly improbable. Even if sales, rents, and employment have increased at particular locations inside these cities, these gains may have been offset by losses elsewhere in the city, resulting in little or no net improvement. Further, as Sternlieb and Hughes have noted, many of the new commercial and office structures support fewer employees for a given amount of floor space than do the older structures they replace, suggesting that employment may have continued to decline even if sales have gone up at particular locations (Sternlieb and Hughes, 1979: 633-634). There is, in brief, no reason to suppose that more recent measurements would show that things have gotten much better.

In sum, there is little cause for optimism about the not-too-distant past of more distressed cities. Over the middle 1970s, people, jobs, and wealth continued to move out of these cities and into newer, more prosperous ones. By almost any reasonable measure of the prosperity of places—levels of population, income, employment, economic activity, and concentration of low-income households—more distressed cities were appreciably worse off in the late 1970s than they were ten years earlier, and more prosperous cities were appreciably better off. Some areas inside some older cities may be reviving, but these revivals do not affect the trend. The urban crisis is not over, nor has moderated so that it is no longer a matter of legitimate governmental concern.

THE QUALITY OF URBAN DATA

Although available data strongly suggest that whatever "revival" has occurred has been limited, it would be premature to conclude that nothing of significance is happening in older cities. This uncertainty is due to a lack of current, reliable information to judge how widespread signs of "revival" are in any given city and whether reviving areas are increasing or shrinking relative to declining ones. We know little about changes in cities, either recently or historically, and it is unlikely that this situation will improve much. Data published by various federal agencies on city conditions are less comprehensive, statistically softer, collected less frequently, and published longer after collection than information collected for counties and states. The federal statistical system for cities is, in short, woefully deficient in many ways.

Four of these dificiencies are especially important. The first, and perhaps most important, is that the federal government collects and publishes less information less frequently on individual cities than it does on individual counties. This is especially true of economic data—employment, income, establishments, payrolls, and a number of other things. The Bureau of Economic Analysis of the Department of Commerce, for example, reports annual income for counties by place of work and place of residence, disaggregated by type of income (wages and salaries, transfer income, and so forth) and by major industrial group, including federal, state, and local government. But the only income data available for cities is the per-capita money income series developed by the Census Bureau. City information is reported only by place of residence and cannot be disaggregated by either type or source of income. No information is available on how much income is earned in cities by those who work there, the relative size of various types of income, or the importance of various industries as sources of income.[3]

This information gap is even larger in reporting economic activity. The Census publishes annually, in the *County Business Patterns* series, information on the numbers of establishments and employees and the size of annual payrolls for 22 major industrial categories. Employment by various local governments, which is not covered in this survey, is reported annually for metropolitan county areas in the *Local Government Employment in Metropolitan Areas* series.

Neither of these series is published for cities. The only systematic sources of information on city economic activity and employment are the Censuses of Business conducted every five years rather than annually—and the industry coverage is not as complete. Governmental employment, reported for metropolitan counties to the county area level, is only available for the city government proper and not for any overlying governments such as school districts or the county. While we know, for example, how many employees of local governments work in Maricopa County, Arizona, we do not know how many of them work in Phoenix. In short, the federal government has no idea how many people work in individual cities at any given time and only makes a partial attempt to find out every five years. Estimates of resident employment are generated regularly as part of the procedure for estimating unemployment statistics, but these figures refer only to the employment of people who *live* in cities, not work in them.

Little information is collected on city economies between censuses. Information on manufacturing value-added and employment is reported yearly in the *Annual Survey of Manufacturing* for cities with more than 20,000 manufacturing employees. Monthly retail sales data are available for five or so of the largest cities. Department store monthly sales are also reported for approximately 40 cities. Apart from these figures, we know next to nothing about what happens in city economies between censuses.

The upshot of this deficiency is that we know a great deal more, and more frequently, about economic conditions in small, rural counties than we do about comparable conditions in some of the nation's largest cities. The best way to illustrate this is by example. Table 3.9 displays the most recent data available on population, income, employment, and economic activity for two places: Washington County, Tennessee, a county with approximately 80,000 people located in northeastern Tennessee, and Chicago, the second largest city in the country. As should be obvious, the federal government collects more information more frequently about Washington County than it does about Chicago. As of December 1979, per-capita income data were available for 1977 for Washington County, and income information disaggregated several ways was available for 1976. But the most current income figure available for Chicago was the money income figure from the 1975 census of business. Similarly, 1978 population data were available for Washington County; the latest population figure for Chicago was for 1977. Finally, a great deal of information on 1977 employment,

(text continued p. 98)

TABLE 3.9 Data for Washington County, Tennessee, and Chicago, Illinois

Statistic	Latest Year Available	Washington County	Latest Year Available	Chicago
Population	1978	83,000	1977[b]	3,062,881
Components of population change [a]				
Total change,	1970-77	8,200	NA	
Births		8,600	NA	
Deaths		5,400	NA	
Net migration		5,100	NA	
Income				
(BEA-place of residence)				
Total	1977[c]	$463m	NA	
Per capita	1977	5,559	NA	
Derivation of personal income by place of residence				
Total labor and proprietors income by place of work	1976[d]	312,854	NA	
Less: personal contributions for social insurance by place of work		17,192	NA	
Net labor and proprietors income by place of work		295,662	NA	
Plus: residence adjustment		1,877	NA	
Net labor and proprietors income by place of residence		297,539	NA	
Plus: dividends, interest & rent [g]		46,933	NA	
Plus: transfer payments		69,854	NA	
Personal income by place of residence		414,326	NA	
Per capita personal income (dollars)		5,094	NA	
Total population (thousands)		81.3	NA	
(BEA-place of work)				
Wage and Salary	1976[d]	$254m	NA	
Other labor	1976	23m	NA	
Proprietors	1976	36m	NA	
farm		8m	NA	
non-farm		28m	NA	
By Industry				
Farm		8,950	NA	
Non-farm		303,904	NA	
Private		243,557	NA	
Ag. Serv., For., Fish., & Other [e]		563	NA	
Mining		0	NA	
Construction		19,841	NA	
Manufacturing		93,967	NA	
Non-durable goods		35,463	NA	
Durable goods		58,504	NA	
Transportation & public utilities		11,490	NA	
Wholesale trade		23,324	NA	
Retail trade		38,771	NA	
Finance, insurance, and real estate		8,769	NA	
Services		46,832	NA	
Government and government enterprises		60,347	NA	
Federal, civilian		26,621	NA	
Federal, military		1,243	NA	
State and local		32,483	NA	
Income (money income)				
(Census-place of residence)				
Per capita	1975[b]	3,913	1975	4,984
Business				
Total Employment** (week of March 12)	1977[e]	24,400	NA	
Agricultural services, forestry, fisheries		31	NA	
Mining		(A)	NA	
Contract construction		1,607	NA	
General contractors and operative builders		634	NA	
General building contractors		508	NA	
Operative builders		70	NA	
Heavy construction contractors		50	NA	
Special trade contractors		923	NA	
Plumbing, heating, air conditioning		209	NA	
Electrical work		150	NA	
Masonry, stonework, and plastering		164	NA	
Masonry and other stonework		(B)	NA	
Plastering, drywall and insulation		86	NA	
Carpentering and flooring		(C)	NA	
Carpentering		(C)	NA	
Roofing and sheet metal work		93	NA	

88 URBAN GOVERNMENT FINANCE

TABLE 3.9 (continued)

Statistic	Latest Year Available	Washington County	Latest Year Available	Chicago
Misc. Special trade contractors		(C)	NA	
Structural steel erection		(B)	NA	
Manufacturing		10,046	NA	
Food and kindred products		742	NA	
Preserved fruits and vegetables		(C)	NA	
Canned fruits and vegetables		(C)	NA	
Grain mill products		(B)	NA	
Flour and other grain mill products		(B)	NA	
Bakery products		(E)	NA	
Bread, cake, and related products		(E)	NA	
Beverages		(C)	NA	
Bottled and canned soft drinks		(C)	NA	
Textile mill products		(G)	NA	
Knitting mills		(B)	NA	
Hosiery, nec		(B)	NA	
Yarn and thread mills		(G)	NA	
Throwing and winding mills		(G)	NA	
Apparel and other textile products		1,114	NA	
Men's and boys' furnishings		(F)	NA	
Men's and boys' separate trousers		(C)	NA	
Men's and boys' working clothing		(E)	NA	
Women's and misses' outerwear		(E)	NA	
Women's and misses' outerwear, nec		(E)	NA	
Women's and children's undergarments		(C)	NA	
Women's and children's underwear		(C)	NA	
Lumber and wood products		(E)	NA	
Sawmill and planing mills		(C)	NA	
Hardwood dimension and flooring		(C)	NA	
Furniture and fixtures		1,182	NA	
Household furniture		1,182	NA	
Wood household furniture		(F)	NA	
Wood TV and radio cabinets		(E)	NA	
Printing and Publishing		263	NA	
Newspapers		(C)	NA	
Chemicals and allied products		(B)	NA	
Rubber and misc. plastics products		(C)	NA	
Fabricated rubber products, nec		(C)	NA	
Stone, clay and glass products		211	NA	
Structural clay products		(C)	NA	
Brick and structural clay tile		(C)	NA	
Concrete, gypsum, and plaster products		(C)	NA	
Concrete products, nec		53	NA	
Primary metal industries		(C)	NA	
Blast furnace and basic steel products		(B)	NA	
Blast furnace and steel mills		(B)	NA	
Nonferrous foundries		(C)	NA	
Nonferrous foundries, nec		(C)	NA	
Fabricated metal products		499	NA	
Cutlery, hand tools, and hardware		286	NA	
Hand and edge tools, nec		286	NA	
Fabricated structural metal products		(C)	NA	
Fabricated structural metal		(C)	NA	
Metal forgings and stampings		(B)	NA	
Automotive stampings		(B)	NA	
Machinery, except electrical		302	NA	
Metalworking machinery		(C)	NA	
Machine tool accessories		(C)	NA	
Misc. machinery, except electrical		(C)	NA	
Machinery, except electrical, nec		(C)	NA	
Electric and electronic equipment		3,014	NA	
Electrical industrial apparatus		(F)	NA	
Industrial controls		(F)	NA	
Household appliances		(G)	NA	
Electric housewares and fans		(F)	NA	
Household appliances, nec		(F)	NA	
Communication equipment		(G)	NA	
Telephone and telegraph apparatus		(G)	NA	
Instruments and related products		(C)	NA	
Medical instruments and supplies		(C)	NA	
Surgical and medical instruments		(C)	NA	
Miscellaneous manufacturing industries		(C)	NA	
Miscellaneous manufacturers		(C)	NA	

TABLE 3.9 (continued)

Statistic	Latest Year Available	Washington County	Latest Year Available	Chicago
Brooms and brushes		(C)	NA	
Administrative and auxilliary		(C)	NA	
Transportation and other public utilities		929	NA	
Local and interurban passenger transit		60	NA	
Trucking and warehousing		373	NA	
Trucking, local and long distance		(E)	NA	
Communication		419	NA	
Telephone communication		(E)	NA	
Radio and television broadcasting		(B)	NA	
Electric, gas, and sanitary services		(B)	NA	
Gas production and distribution		(B)	NA	
Wholesale trade		1,775	NA	
Wholesale trade – durable goods		792	NA	
Motor vehicles & automotive equipment		118	NA	
Automotive parts and supplies		102	NA	
Furniture and home furnishings		95	NA	
Home furnishings		(B)	NA	
Hardware, plumbing & heating equipment		210	NA	
Hardware		(C)	NA	
Plumbing and hydronic heating supplies		(B)	NA	
Machinery, equipment, and supplies		286	NA	
Construction and mining machinery		61	NA	
Industrial machinery and equipment		75	NA	
Wholesale trade – nondurable goods		963	NA	
Drugs, proprietaries and sundries		(B)	NA	
Groceries and related products		534	NA	
Groceries, general line		(E)	NA	
Dairy products		(B)	NA	
Meats and meat products		52	NA	
Farm-product rawmaterieals		93	NA	
Farm-product raw materials, nec		68	NA	
Petroleum and petroleum products		94	NA	
Petroleum bulk stations & terminals		94	NA	
Beer, wine, and distilled beverages		93	NA	
Beer and ale		93	NA	
Miscellaneous nondurable goods		73	NA	
Retail trade		4,863	(see Retail Trade below)	
Building materials and garden supplies		275	NA	
Lumber and other building materials		157	NA	
General merchandise stores		807	NA	
Department stores		597	NA	
Miscellaneous general merchandise stores		(C)	NA	
Food stores		684	NA	
Grocery stores		648	NA	
Automotive dealers & service stations		876	NA	
New and used car dealers		405	NA	
Used car dealers		58	NA	
Auto and home supply stores		189	NA	
Gasoline service stations		208	NA	
Apparel and economy stores		(E)	NA	
Women's ready-to-wear stores		117	NA	
Shoe stores		63	NA	
Furniture and home furnishings stores		224	NA	
Furniture and home furnishings stores		137	NA	
Furniture stores		109	NA	
Radio, television and music stores		67	NA	
Eating and drinking places		991	NA	
Eating places		867	NA	
Drinking places		63	NA	
Miscellaneous retail		548	NA	
Drug stores and proprietary stores		156	NA	
Liquor stores		100	NA	
Miscellaneous shopping goods stores		156	NA	
Retail stores, nec		77	NA	
Administrative and auxilliary		(C)	NA	
Finance, insurance, and real estate		1,066	NA	
Banking		407	NA	
Commercial and stock savings bank		407	NA	
Credit agencies other than banks		174	NA	
Savings and loan associations		74	NA	
Personal credit institutions		89	NA	
Insurance carriers		181	NA	

90 URBAN GOVERNMENT FINANCE

TABLE 3.9 (continued)

Statistic	Latest Year Available	Washington County	Latest Year Available	Chicago
Life insurance		160	NA	
Insurance agents, brokers & service		136	NA	
Real estate		131	NA	
Real estate operators and leasors		80	NA	
Services		4,126	NA	(See Service Industries Below)
Hotels and other lodging places		136	NA	
Hotels, motels and tourist courts		(C)	NA	
Personal services		378	NA	
Laundry, cleaning and garment services		88	NA	
Dry cleaning plants except rug		71	NA	
Beauty shops		140	NA	
Funeral service and crematories		53	NA	
Miscellaneius personal services		(B)	NA	
Business services		100	NA	
Services to buildings		57	NA	
Auto repair, services, and garages		190	NA	
Automotive repair shops		155	NA	
General automotive repair shops		52	NA	
Automotive repair shops, nec		83	NA	
Miscellaneous repair services		60	NA	
Amusement and recreation services		140	NA	
Misc. amusement, recreational services		(B)	NA	
Health services		2,138	NA	
Offices of physicians		200	NA	
Offices of dentists		100	NA	
Nursing and personal care facilities		175	NA	
Hospitals		1,423	NA	
Medical and dental laboratories		52	NA	
Health, and allied services, nec		(B)	NA	
Legal services		84	NA	
Educational services		72	NA	
Social services		244	NA	
Social services, nec		(C)	NA	
Residential care		(C)	NA	
Membership organizations		363	NA	
Civic and social associations		78	NA	
Religious organizations		186	NA	
Miscellaneous services		128	NA	
Engineering & architectural services		71	NA	
Accounting, auditing & bookkeeping		54	NA	
Nonclassifiable establishments		(B)	NA	
Manufacturing[f]				
Employees	1977	11,700	1976[g]	381,600
Production workers		8,200		245,700
Retail Trade[h] (week of March 12)				
Paid employees	1977	4,795	1977[i]	178,336
Building materials, etc.		283		3,337
General merchandise		800		28,972
Food stores		677		23,096
Automotive dealers		632		7,806
Gasoline service stations		238		7,209
Apparel and Acc. stores		315		14,086
Furniture, equipment, etc.		226		5,703
Eating and drinking		1,014		50,109
Drug and proprietary		146		8,395
Miscellaneous retail stores		391		28,147
Service Industries[j] (week of March 12)				
Paid employees	1977	1,271	1977[k]	158,015
Hotels, motels, etc.		188		18,086
Personal services		389		16,062
Business services		162		74,313
Auto. repair, srevices, etc.		156		8,947
Miscellaneous repair services		80		4,719
Amusement and recreation		105		6,887
Dental		withheld		700
Legal services		84		14,676
Engineering, etc.		withheld		13,623

TABLE 3.9 (continued)

Statistic	Latest Year Available	Washington County	Latest Year Available	Chicago
Annual Payroll**	1977e	$213m	NA	
Agricultural services, forestry, fisheries		186	NA	
Mining		(D)	NA	
Contract construction		16,415	NA	
General contractors and operative builders		5,491	NA	
General building contractors		4,571	NA	
Operative builders		499	NA	
Heavy construction contractors		613	NA	
Special trade contractors		10,311	NA	
Plumbing, heating, air conditioning		1,794	NA	
Electrical work		2,166	NA	
Masonry, stonework, and plastering		1,138	NA	
Masonry and other stonework		(D)	NA	
Plastering, drywall and insulation		631	NA	
Carpentering and flooring		(D)	NA	
Carpentering		(D)	NA	
Roofing and sheet metal work		848	NA	
Misc. special trade contractors		(D)	NA	
Structural steel erection		(D)	NA	
Manufacturing		93,600	NA	
Food and kindred products		8,449	NA	
Preserved fruits and vegetables		(D)	NA	
Canned fruits and vegetables		(D)	NA	
Grain mill products		(D)	NA	
Flour and other grain mill products		(D)	NA	
Bakery products		(D)	NA	
Bread, cake, and related products		(D)	NA	
Beverages		(D)	NA	
Bottled and canned soft drinks		(D)	NA	
Textile mill products		(D)	NA	
Knitting mills		(D)	NA	
Hosiery, nec		(D)	NA	
Yarn and thread mills		(D)	NA	
Throwing and winding mills		(D)	NA	
Apparel and other textile products		7,274	NA	
Men's and boys' furnishings		(D)	NA	
Men's and boys' separate trousers		(D)	NA	
Men's and boys' working clothing		(D)	NA	
Women's and misses' outerwear		(D)	NA	
Women's and misses' outerwear, nec		(D)	NA	
Women's and children's undergarments		(D)	NA	
Women's and children's underwear		(D)	NA	
Lumber and wood products		(D)	NA	
Sawmill and planing mills		(D)	NA	
Hardwood dimension and flooring		(D)	NA	
Furniture and fixtures		8,455	NA	
Household furniture		8,455	NA	
Wood household furniture		(D)	NA	
Wood TV and radio cabinets		(D)	NA	
Printing and publishing		2,062	NA	
Newspapers		(D)	NA	
Chemicals and allied products		(D)	NA	
Rubber and misc. plastics products		(D)	NA	
Fabricated rubber products, nec		(D)	NA	
Stone, clay and glass products		2,109	NA	
Structural clay products		(D)	NA	
Brick and structural clay tile		(D)	NA	
Concrete, gypsum, and plaster products		(D)	NA	
Concrete products, nec		507	NA	
Primary metal industries		(D)	NA	
Blast furnace and basic steel products		(D)	NA	
Blast furnace and steel mills		(D)	NA	
Nonferrous foundries		(D)	NA	
Nonferrous foundries, nec		(D)	NA	
Fabricated metal products		4,606	NA	
Cutlery, hand tools, and hardware		2,465	NA	
Hand and edge tools, nec		2,465	NA	
Fabricated structural metal products		(D)	NA	
Fabricated structural metal		(D)	NA	
Metal forgings and stampings		(D)	NA	
Automotive stampings		(D)	NA	
Machinery, except electrical		3,140	NA	
Metalworking machinery		(D)	NA	
Machine tool accessories		(D)	NA	

92 URBAN GOVERNMENT FINANCE

TABLE 3.9 (continued)

Statistic	Latest Year Available Washington County	Latest Year Available Chicago
Misc. machinery, except electrical	(D)	NA
Machinery, except electrical, nec	(D)	NA
Electric and electronic equipment	31,944	NA
Electrical industrial apparatus	(D)	NA
Industrial controls	(D)	NA
Household appliances	(D)	NA
Electric housewares and fans	(D)	NA
Household appliances, nec	(D)	NA
Electric housewares and fans	(D)	NA
Household appliances, nec	(D)	NA
Communication equipment	(D)	NA
Telephone and telegraph apparatus	(D)	NA
Instruments and related products	(D)	NA
Medical instruments and supplies	(D)	NA
Surgical and medical instruments	(D)	NA
Miscellaneous manufacturing industries	(D)	NA
Miscellaneous manufacturers	(D)	NA
Brooms and brushes	(D)	NA
Administrative and auxilliary	(D)	NA
Transportation and other public utilities	10,863	NA
Local and interurban passenger transit	366	NA
Trucking and warehousing	5,340	NA
Trucking, local and long distance	(D)	NA
Communication	4,446	NA
Telephone communication	(D)	NA
Radio and television broadcasting	(D)	NA
Electric, gas, and sanitary services	(D)	NA
Gas production and distribution	(D)	NA
Wholesale trade	18,729	NA
Wholesale trade–durable goods	8,765	NA
Motor vehicles & automotive equipment	1,215	NA
Automotive parts and supplies	1,013	NA
Furniture and home furnishings	848	NA
Home furnishings	(D)	NA
Hardware, plumbing & heating equipment	2,663	NA
Hardware	(D)	NA
Plumbing and hydronic heating supplies	(D)	NA
Machinery, equipment, and supplies	3,145	NA
Construction and mining machinery	823	NA
Industrial machinery and equipment	655	NA
Wholesale trade–nondurable goods	9,964	NA
Drugs, proprietaries and sundries	(D)	NA
Groceries and related products	5,353	NA
Groceries, general line	(D)	NA
Dairy products	(D)	NA
Meats and meat products	405	NA
Farm-product raw materials	286	NA
Farm-product raw materials, nec	188	NA
Petroleum and petroleum products	1,108	NA
Petroleum bulk stations & terminals	1,102	NA
Beer, wine, and distilled beverages	1,450	NA
Beer and ale	1,450	NA
Miscellaneous nondurable goods	734	NA
Retail trade	32,898	NA
Building materials and garden supplies	2,446	NA
Lumber and other building materials	1,566	NA
General merchandise stores	5,503	NA
Department stores	4,584	NA
Miscellaneous general merchandise stores	(D)	NA
Food stores	4,294	NA
Grocery stores	4,149	NA
Automotive dealers & service stations	9,255	NA
New and used car dealers	4,792	NA
Used car dealers	642	NA
Auto and home supply stores	1,702	NA
Gasoline service stations	1,039	NA
Apparel and economy stores	(D)	NA
Women's ready-to-wear stores	594	NA
Shoe stores	361	NA
Furniture and home furnishings stores	1,609	NA
Furniture and home furnishings stores	1,138	NA
Furniture stores	910	NA

TABLE 3.9 (continued)

Statistic	Latest Year Available	Washington County	Latest Year Available	Chicago
Radio, television and music stores		384	NA	
Eating and drinking places		4,671	NA	
Eating places		4,211	NA	
Drinking places		339	NA	
Miscellaneous retail		3,069	NA	
Drug stores and proprietary stores		972	NA	
Liquor stores		643	NA	
Miscellaneous shopping goods stores		851	NA	
Retail stores, nec		261	NA	
Administrative and auxilliary		(D)	NA	
Finance, insurance, and real estate		9,930	NA	
Banking		3,311	NA	
Commercial and stock savings bank		3,311	NA	
Credit agencies other than banks		1,894	NA	
Savings and loan associations		829	NA	
Personal credit institutions		965	NA	
Insurance carriers		2,181	NA	
Life insurance		2,065	NA	
Insurance agents, brokers & service		1,344	NA	
Real estate		717	NA	
Real estate operators and leasors		411	NA	
Services		30,206	NA	(See Service Industries Below)
Hotels and other lodging places		671	NA	
Hotels, motels and tourist courts		(D)	NA	
Personal services		1,855	NA	
Laundry, cleaning and garment services		368	NA	
Dry cleaning plants except rug		295	NA	
Beauty shops		715	NA	
Funeral service and crematories		400	NA	
Miscellaneous personal services		(D)	NA	
Business services		1,053	NA	
Services to buildings		402	NA	
Auto repair, services, and garages		1,436	NA	
Automotive repair shops		1,316	NA	
General automotive repair shops		319	NA	
Automotive repair shops, nec		638	NA	
Miscellaneous repair services		566	NA	
Amusement and recreation services		683	NA	
Misc. amusement, recreational services		(D)	NA	
Health services		18,582	NA	
Offices of physicians		4,594	NA	
Offices of dentists		1,038	NA	
Nursing and personal care facilities		1,063	NA	
Hospitals		10,864	NA	
Medical and dental laboratories		446	NA	
Health, and allied services, nec		(D)	NA	
Legal services		639	NA	
Educational services		516	NA	
Social services		1,077	NA	
Social services, nec		(D)	NA	
Residential care		(D)	NA	
Membership organizations		1,612	NA	
Civic and social associations		510	NA	
Religious organizations		712	NA	
Miscellaneous services		1,391	NA	
Engineering & architectural services		976	NA	
Accounting, auditing & bookkeeping		404	NA	
Nonclassifiable establishments		(D)	NA	
Manufacturing[j] (millions of dollars)				
Payroll	1977	109.9	1976[g]	5,154
Production wages		61.9		2,695
Retail Trade[k] (millions of dollars)				
Payroll	1977	31.5	1979[i]	1,292
Building materials, etc.		2.3		22.5
General merchandise		5.7		193.9
Food stores		4.2		193.1
Automotive dealers		7.0		112.2
Gas. service stations		1.2		40.5

94 URBAN GOVERNMENT FINANCE

TABLE 3.9 (continued)

Statistic	Latest Year Available	Washington County	Latest Year Available	Chicago
Population	1978	83,000	1977[b]	3,062,881
Apparel and acc. stores		1.7		106.3
Furniture, equipment, etc.		1.7		61.5
Eating and drinking		4.6		238.2
Drug and proprietary		1.0		54.7
Miscellaneous retail stores		2.2		256.3
Service Industries[j] (millions of dollars)				
Payroll	1977	8.2	1977[k]	1,813
Hotels, motels, etc.		.9		120.6
Personal services		1.9		166.9
Business services		1.1		816.0
Auto. repair, services, etc.		1.2		91.0
Miscellaneous repair services		.6		60.0
Amusement and recreation		.4		73.4
Dental		withheld		7.5
Legal services		.8		254.0
Engineering, etc.		withheld		273.4
Number of Establishments	1977[e]	1,639		NA
Agricultural services, forestry, fisheries		12		NA
Mining		1		NA
Contract construction		220		NA
General contractors and operative builders		76		NA
General building contractors		44		NA
Operative builders		12		NA
Heavy construction contractors		10		NA
Special trade contractors		134		NA
Plumbing, heating, air conditioning		28		NA
Electrical work		18		NA
Masonry, stonework, and plastering		34		NA
Masonry and other stonework		22		NA
Plastering, drywall and insulation		9		NA
Carpentering and flooring		11		NA
Carpentering		9		NA
Roofing and sheet metal work		11		NA
Misc. special trade contractors		11		NA
Structural steel erection		1		NA
Manufacturing		103		NA
Food and kindred products		13		NA
Preserved fruits and vegetables		1		NA
Canned fruits and vegetables		1		NA
Grain mill products		2		NA
Flour and other grain mill products		1		NA
Bakery products		2		NA
Bread, cake, and related products		2		NA
Beverages		3		NA
Bottled and canned soft drinks		3		NA
Textile mill products		4		NA
Knitting mills		1		NA
Hosiery, nec		1		NA
Yarn and thread mills		2		NA
Throwing and winding mills		2		NA
Apparel and other textile products		7		NA
Men's and boys' furnishings		3		NA
Men's and boys' separate trousers		1		NA
Men's and boys' working clothing		2		NA
Women's and misses' outerwear		1		NA
Women's and misses' outerwear, nec		1		NA
Women's and children's undergarments		1		NA
Women's and children's underwear		1		NA
Lumber and wood products		11		NA
Sawmill and planing mills		5		NA
Hardwood dimension and flooring		1		NA
Furniture and fixtures		7		NA
Household furniture		7		NA
Wood household furniture		2		NA
Wood TV and radio cabinets		1		NA
Printing and publishing		10		NA
Newspapers		2		NA
Chemicals and allied products		3		NA
Rubber and misc. plastics products		4		NA
Fabricated rubber products, nec		2		NA
Stone, clay and glass products		8		NA
Structural clay products		1		NA

TABLE 3.9 (continued)

Statistic	Latest Year Available	Washington County	Latest Year Available	Chicago
Brick and structural clay tile		1		NA
Concrete, gypsum, and plaster products		6		NA
Concrete products, nec		3		NA
Primary metal industries		2		NA
Blast furnace and basic steel products		1		NA
Blast furnace and steel mills		1		NA
Nonferrous foundries		1		NA
Nonferrous foundries, nec		1		NA
Fabricated metal products		9		NA
Cutlery, hand tools, and hardware		3		NA
Hand and edge tools, nec		3		NA
Fabricated structural metal products		2		NA
Fabricated structural metal		1		NA
Metal forgings and stampings		2		NA
Automotive stampings		1		NA
Machinery, except electrical		7		NA
Metalworking machinery		2		NA
Machine tool accessories		1		NA
Misc. machinery, except electrical		4		NA
Machinery, except electrical, nec		4		NA
Electric and electronic equipment		6		NA
Electrical industrial apparatus		1		NA
Industrial controls		1		NA
Household appliances		3		NA
Electric housewares and fans		2		NA
Household appliances, nec		1		NA
Communication equipment		1		NA
Telephone and telegraph apparatus		1		NA
Instruments and related products		3		NA
Medical instruments and supplies		2		NA
Surgical and medical instruments		2		NA
Miscellaneous manufacturing industries		4		NA
Miscellaneous manufacturers		3		NA
Brooms and brushes		1		NA
Administrative and auxilliary		3		NA
Transportation and other public utilities		43		NA
Local and interurban passenger transit		3		NA
Trucking and warehousing		28		NA
Trucking, local and long distance		26		NA
Communication		7		NA
Telephone communication		3		NA
Radio and television broadcasting		2		NA
Electric, gas, and sanitary services		3		NA
Gas production and distribution		1		NA
Wholesale trade		136		NA
Wholesale trade—durable goods		79		NA
Motor vehicles & automotive equipment		16		NA
Automotive parts and supplies		11		NA
Furniture and home furnishings		7		NA
Home furnishings		3		NA
Hardware, plumbing & heating supplies		7		NA
Hardware		2		NA
Plumbing and hydronic heating supplies		2		NA
Machinery, equipment, and supplies		33		NA
Construction and mining machinery		4		NA
Industrial machinery and equipment		4		NA
Wholesale trade—nondurable goods		59		NA
Drugs, proprietaries and sundries		2		NA
Groceries and related products		19		NA
Groceries, general line		5		NA
Dairy products		3		NA
Meats and meat products		6		NA
Farm-product raw materials		8		NA
Farm-product raw materials, nec		5		NA
Petroleum and petroleum products		10		NA
Petroleum bulk stations & terminals		9		NA
Beer, wine, and distilled beverages		5		NA
Beer and ale		5		NA
Miscellaneous nondurable goods		8		NA
Retail trade		482		NA
Building materials and garden supplies		27		NA
Lumber and other building materials		11		NA
General merchandise stores		16		NA

TABLE 3.9 (continued)

Statistic	Latest Year Available	Washington County	Latest Year Available	Chicago
Department stores		6	NA	
Miscellaneous general merchandise stores		5	NA	
Food stores		58	NA	
Grocery stores		51	NA	
Automotive dealers & service stations		100	NA	
New and used car dealers		13	NA	
Used car dealers		12	NA	
Auto and home supply stores		13	NA	
Gasoline service stations		57	NA	
Apparel and economy stores		44	NA	
Women's ready-to-wear stores		14	NA	
Shoe stores		13	NA	
Furniture and home furnishings stores		37	NA	
Furniture and home furnishings stores		21	NA	
Furniture stores		15	NA	
Radio, television and music stores		10	NA	
Eating and drinking places		88	NA	
Eating places		72	NA	
Drinking places		13	NA	
Miscellaneous retail		104	NA	
Drug stores and proprietary stores		16	NA	
Liquor stores		21	NA	
Miscellaneous shopping goods stores		32	NA	
Retail stores, nec		24	NA	
Administrative and auxilliary		8	NA	
Finance, insurance, and real estate		182	NA	
Banking		27	NA	
Commercial and stock savings bank		27	NA	
Credit agencies other than banks		28	NA	
Savings and loan associations		5	NA	
Personal credit institutions		21	NA	
Insurance carriers		17	NA	
Life insurance		14	NA	
Insurance agents, brokers & service		39	NA	
Real estate		57	NA	
Real estate operators and leasors		42	NA	
Services		440	NA	(See Service Industries Below)
Hotels and other lodging places		7	NA	
Hotels, motels and tourist courts		6	NA	
Personal services		77	NA	
Laundry, cleaning and garment services		19	NA	
Dry cleaning plants except rug		9	NA	
Beauty shops		41	NA	
Funeral service and crematories		6	NA	
Miscellaneous personal services		4	NA	
Business services		28	NA	
Services to buildings		7	NA	
Auto repair, services, and garages		40	NA	
Automotive repair shops		34	NA	
General automotive repair shops		14	NA	
Automotive repair shops, nec		11	NA	
Miscellaneous repair services		21	NA	
Amusement and recreation services		18	NA	
Misc. amusement, recreational services		6	NA	
Health services		103	NA	
Offices of physicians		46	NA	
Offices of dentists		33	NA	
Nursing and personal care facilities		5	NA	
Hospitals		4	NA	
Medical and dental laboratories		4	NA	
Health, and allied services, nec		3	NA	
Legal services		33	NA	
Educational services		6	NA	
Social services		12	NA	
Social services, nec		10	NA	
Residential care		2	NA	
Membership organizations		64	NA	
Civic and social associations		13	NA	
Religious organizations		27	NA	
Miscellaneous services		26	NA	
Engineering & architectural services		16	NA	
Accounting, auditing & bookkeeping		11	NA	

TABLE 3.9 (continued)

Statistic	Latest Year Available	Washington County	Latest Year Available	Chicago
Nonclassifiable establishments		18	NA	
Manufacturing [f]				
Establishments	1977	113	NA	
Retail Trade [h]				
Establishments	1977	738	1977[i]	20,256
Sole proprietorships		425		10,418
Partnerships		35		1,617
Establishments with payroll		491		13,502
Building materials, etc.		31		386
General merchandise		14		286
Food stores		56		1,625
Automotive dealers		43		462
Gas. service stations		66		1,095
Apparel and acc. stores		45		1,509
Furniture, equipment etc.		41		798
Eating and drinking		89		4,145
Drug and proprietary		15		800
Misc. retail stores		89		2,418
Service Industries [j]				
Establishments	1977	583	1977[k]	24,834
Sole proprietorships		442		16,665
Partnerships		49		2,033
Establishments with payroll		254		10,738
Hotels, motels, etc.		7		224
Personal services		78		1,959
Business services		32		3,501
Auto. repair, services, etc.		38		1,226
Misc. repair services		23		564
Amusement and recreation		22		577
Dental		3		96
Legal services		33		2,194
Engineering, etc.		18		15.4
Manufacturing Data [j] (millions of dollars & hours) (general)				
Man-hours	1977	15.4	1976[g]	475.1
Value added		$243.9		$9,863
Cost of materials		244.8		11,302
Value of shipments		488.7		21,121
New capital expenditures		12.8		459,000
End-of-year inventories	1976[g]	62		2,745
Sales Data (millions of dollars)				
Retail Trade [h]				
Receipts (all estab)	1977	281.0	1979[l]	10,413
Building materials, etc.		28.4	1977[i]	231.3
General Merchandise		37.8	1979[l]	2,658
Food stores		53.4		1,755
Automotive dealers		70.9		1,216
Gas. service stations		19.0		542.9
Apparel and acc. stores		13.0		672.9
Furniture, equipment, etc.		13.4		440.1
Eating and drinking		18.4		1,018
Drug and proprietary		6.7		392.3
Misc. retail stores		19.9		1,561
Service Industries [j]				
Receipts (all estab)	1977	29.1	1977[k]	5,121
Hotels, motels, etc.		3.5		369.0
Personal services		5.6		368.0
Business services		3.4		2,018
Auto. repair, services, etc.		5.7		392.8
Misc. repair services		2.3		178.1
Amusement and recreation		withheld		347.9
Dental		withheld		18.7
Legal services		3.3		881.6
Engineering, etc.		2.3		547.1

KEY: A: 0-19; B: 20-99; C: 100-249; E: 250-499; F: 500-999; G: 1,000-2,499; H: 2,500-4,999; I: 5,000-9,999; J: 10,000-24,999; K: 25,000-40,999; L: 50,000-99,999; M: 100,000 or more

(Table 3.9 notes on p. 98)

NOTE: Table 3.9 presents the most current data available on a regular basis from the federal statistical system for the four areas of population, income, business and employment. Decennial census material from 1970, because of its age, is excluded. Also excluded are the various state and private sources of local area statistics which, while useful in specific instances, are dissimilar in method, making comparative analysis between localities and sources impossible. Data are current as of December 1979.

[a] *Current Population Reports, Federal-State Cooperative Program for Population Estimates. Estimates of the Population of Tennessee Counties and Metropolitan Areas, July 1, 1977 (Revised) and 1978 (Provisional).* August 1979. (Commerce-Bureau of the Census)

[b] *General Revenue Sharing Initial State and Local Data Elements Entitlement Period II.* August 1979. (Treasury-Office of Revenue Sharing)

[c] *Survey of Current Business.* April 1979. (Commerce-Bureau of Economic Analysis)

[d] *Local Area Personal Income, 1971-1976.* August 1978. (Commerce-Bureau of Economic Analysis)

[e] *County Business Patterns.* June 1979. (Commerce-Bureau of the Census)

[f] *1977 Census of Manufactures* (Preliminary Report). Geographic Area Series. Oct. 1979. (Commerce-Bureau of the Census)

[g] *Annual Survey of Manufactures. Statistics for States, Standard Metropolitan Statistical Areas, Large Industrial Counties and Selected Cities.* February 1978. (Commerce-Bureau of the Census)

[h] *1977 Census of Retail Trade.* Geographic Area Series. Tennessee. August 1979. (Commerce-Bureau of the Census)

[i] *1977 Census of Retail Trade.* Geographic Area Series. Illinois. June 1979. (Commerce-Bureau of the Census)

[j] *1977 Census of Service Industries.* Geographic Area Series. Tennessee. Jan. 1980. (Commerce-Bureau of the Census)

[k] *1977 Census of Service Industries.* Geographic Area Series. Illinois. Jan. 1980. (Commerce-Bureau of the Census)

[l] *Current Business Reports. Monthly Retail Trade.* Feb. 1980. (Commerce-Bureau of the Census)

establishments, and payrolls was available for the county but not for the city when these figures were gathered. Although some of these figures will become available for Chicago when the full reports from the 1977 censuses of business are released, it is worth noting that the 1977 county figures have been out, as of this writing, for almost a year; and 1978 figures were available in the spring of 1980. The next installment of information on Chicago's economic conditions, with the exceptions noted above, will not be available until results from the 1982 census of business are released.

The second conclusion about the relative merits of city and county data is that many of the city numbers are softer than comparable county data. The procedures used to construct figures for cities rely more heavily on imputations, interpolations, and prorationing than do the procedures by which comparable county figures are constructed. Many of the data elements required to construct the figures of interest to students of city conditions are available for counties but not for cities. Accordingly, city figures must be estimated from county data rather than being measured directly. In many cases, this estimation is done on the basis of decennial census relationships, which means that splits between a city and the balance of a county are made on the basis of *relations* that may be as much as ten years old. We attempt no judgment on the extent or direction of the bias introduced by this sort of procedure; but data developed without using these sorts of techniques are preferable to those requiring them.

This argument is best made by example. County population estimates are averages of three separate methods. All three rely on vital statistics—records of births and deaths—universally available for counties. A variety of information is used to estimate net migration. The resulting estimates are "hard" enough to produce not only estimates of total population, but also estimates of population change components—births, deaths, and migrations—published with the population figures.

But city population is estimated by a single procedure that relies less heavily on city-specific data. Because vital statistics are not generally available for cities, county births and deaths must be prorated down to individual cities, using relations derived from the 1970 census. Net migration is estimated using only one of the sources available for counties—a series of tabulations from Internal Revenue Service (IRS) data derived from individual tax returns. Estimates of the components of population change are not published for cities.

Similar procedures, again relying on benchmarks from the 1970 census, are used to estimate residential employment and unemployment in cities. The entire set of procedures used to calculate these rates has come under considerable attack.[5] We do not wish to join this onslaught except to note that, with the exception of a few cities for which current unemployment claims data are available or for which unemployment can be estimated more directly from information gathered from Current Population Survey figures, city employment and unemployment figures are constructed by splitting various county figures on the basis of various relations drawn from the 1970 census. Although the county figures are themselves partly constructed by similar procedures, county claims and establishment employment information, used to construct rates, are available fairly quickly and for almost all counties in the nation. In short, the county rates can be judged to be "harder" than the city rates.

A slightly different prorationing technique is used to calculate city per-capita income figures. Similar procedures are used to estimate city and county income for several different sources of income, but city-level information is generally not available for payments made by a number of income transfer programs such as Social Security, Aid to Families with Dependent Children (AFDC), general assistance, and various federal and military pension programs. Information on payments under these programs, amounting to as much as 30% of total income in some areas, is usually available at the county level. Income from these programs is prorated to cities under the assumption that city transfer payments have grown at the same rate as county payments.

A third shortcoming is that much useful information is not collected at all for individual cities except during the decennial census. For example, information on the numbers of households of various sizes, types, and income levels, and the numbers and relative concentration of the poverty population are only available every ten years for individual cities. Although some of this information is reported for cities in the aggregate approximately every three years in the *Characteristics of the Metropolitan and Nonmetropolitan Population* series, these figures are of relatively little help in attempting to gauge changes in individual cities.

A fourth difficulty with the existing urban statistical system is that it takes a long time to produce those data that it does collect. Most of the series we have referred to take at least two years to produce; for several series, the lag between collection and publication is even longer. As of March 1980, reports for cities were not yet available from the 1977 Census of Manufacturing. The most recent information available on city population is three years old; the most current income data for cities are five years old. Information on city government finances and employment is just now coming out for the 1978 fiscal year. The only series for which reasonably current figures are available is residential employment and unemployment, which is available through 1979 for cities that serve as prime sponsors of the Department of Labor's Comprehensive Education and Training (CETA) program. In short, the most current diagnosis in any detail we can offer of conditions in particular cities is from two to five years old; for some important particulars, our information dates from 1970.

The final, and perhaps most pessimistic, conclusion is that conditions are unlikely to improve. The Census Bureau conducts a cooperative program with state governments to improve the quality of county population estimates; *no* comparable effort is underway for cities. Perhaps, more importantly, planning for the 1985 mid-decade census calls only for preparing detailed population characteristics down to the individual SMSA, rather than the city, level. This decision not only means that much information of interest to students of cities will continue to be available only every ten years, but also that intercensal data will continue to be generated from benchmarks that may be out-dated and misleading. The Department of Labor (1980), in response to criticisms levied by the Levitan Commission, has announced only limited changes in the way unemployment data are generated. These decisions, among others, together with the problems outlined above, lead us to echo Senator Moynihan's complaint:

> All the numbers I have used here are official; but let me say right off that they are incomplete. We haven't anything like the data base we want. For three years I have been talking to Cabinet officers about this, asking their departments' help. (You never do anything about a problem in Washington until you learn to measure it.) I have been met with incomprehension and near total failure to respond. It is as if someone called at the Labor Department in 1934 suggesting that the Bureau of Labor Statistics begin measuring unemployment on a regular basis . . . only to have Madame Perkins ask, "Why on earth would anyone want to keep count of the number of persons out of work?" [1980].

CONCLUSIONS—THE URBAN RESEARCH AGENDA

The earlier parts of this chapter examined the extent of urban "revitalization" and the limitations of existing information about trends and conditions in cities. Our conclusions on both questions have been negative. Available evidence shows relatively few signs of incipient revival in any appreciable number of older cities, but the data are old enough and flawed enough to make any conclusions shaky.

In this section we suggest some ways to examine the changes allegedly now occurring in many cities. In our view, the next major round of research into urban conditions should focus on the information collected in the 1980 census, which will begin to be available at the city level in late 1981 and at the subcity level in early 1982. There are three lines of potentially productive investigation.

The first is frankly retrospective. The scenario outlined in the first section about the demographic shift allegedly under way in cities needs to be tested and its significance assessed for individual places. As noted earlier, aggregate changes in either the number of households or income levels were not big enough to indicate widespread residential revival, but the picture for individual cities might be more favorable. In any case, considerable descriptive research needs to be done on shifts in household formation and migration patterns in individual cities over the 1970s.

Second, researchers should examine the relative size of areas inside cities that have "revived" since 1970. The 1980 data provides an opportunity to disaggregate below the city level and develop taxonomies of urban territory to judge the significance of residential revival. Block group or tract data could identify areas where median income and housing values have gone up and where changes have occurred in social and demographic composition. These might then be compared in size and population to areas that have declined by similar criteria. This type of analysis would allow an assessment, on a better empirical basis, of the size and significance of residential revival in individual cities.

Third, urban researchers should assess changes in the social and economic disparity between central cities and their suburbs since 1970. In an earlier article, Nathan and Adams (1976) found that older metropolitan areas manifested the classic "doughnut" pattern of a poor core city surrounded by wealthy suburbs. Repeating this analysis with 1980 data might show some reductions in this pattern in some places; perhaps prosperous groups are no longer moving out as much, and poorer groups are moving out more. Several observers have noted a recent increase in the suburbanization of minority populations in some cities, resulting in the formation of poorer suburbs. In addition, 1980 data may show that more prosperous cities, which appeared well off relative to their suburbs in 1970, have lost ground relative to their suburbs over the seventies due to a decline in annexation and an increased concentration of lower income groups.

NOTES

1. A number of objections have been lodged against this index, which was initially developed by Paul R. Dommel. First, it has been argued that the use of per-capita income understates the spread between rich and poor cities relative to

other measures such as the percentage of the population below the poverty level. It has also been argued that figures fail to reflect regional variations in the cost of living, and should be adjusted for price differences between cities. We agree with both these objections in principle, but lack the information to accommodate them. There are no data available on how many poor people lived in individual cities in 1960 or on price levels in all the cities surveyed here over the past twenty years. A further objection has been lodged against the use of older housing, on the ground that this figure does not reflect levels of housing rehabilitation and is hence implicitly biased towards new construction as a measure of city health. This objection is, in our view, misguided. The fraction of older housing is intended as a proxy for the age of city capital stock, which includes such items as streets and sewers, rather than a measure of the conditions of a city's housing. Cities with high fractions of older housing might be expected to also have an older capital stock.

2. Legal services, dental labs, engineering, architectural and land surveying were added to the 1972 Census of Selected Services. These additions account for approximately 15 percent of 1972 service receipts for most cities. Further coverage changes were made in the 1977 Census of Selected Services. In particular, coverage of tax-exempt establishments has been expanded considerably and will be reported separately from taxable activities. No such distinction was made in the 1972 Census. Since complete data on tax-exempt establishments were not available at the time this article was prepared, no attempt has been made to adjust the 1977 figures for coverage changes. For further information on the comparability question, see the appendexes to *U.S. Census of Selected Services: Volume VII: Area Statistics*, 1972 and 1977.

3. The BEA and Census use different definitions of what constitutes income, causing their estimates of per-capita income for a particular place to differ. The BEA figure is generally higher, since it includes several varieties of imputed income, such as rent, which are excluded from the Census figure.

4. A word to the suspicious: Neither of these places was chosen because of statistical idiosyncracies that increases the available data for the county and decreases it for the city. If we had chosen to play games with the numbers, we would have chosen one of the numerous large cities, such as Newark, Phoenix, or Buffalo, which are not covered in the monthly retail sales series, and would have picked a metropolitan county so that we could list more recent county area government employment data, to name but two examples. Both of these places were chosen more or less at random.

5. For a complete description and a lengthy critique of procedures used to estimate these rates, see Presidential Commission on Employment and Unemployment Statistics, *Counting the Labor Force* (Washington, 1979). Several studies, relying on survey results, have argued that current procedures severely underestimate urban unemployment. See, for example, "Surveying Unemployment in Cleveland," (Planning and Research Staff, Western Resource Area Manpower Consortium; November 1977) and E. Terrence Jones, *et.al.*, "Measuring Unemployment in the City of St. Louis," (University of Missouri; processed, 1975).

REFERENCES

ALLMAN, T. D. (1978) "The urban crisis leaves town." Harpers (December).
BARABBA, V. (1980) "The demographic future of the cities of America." Presented at the National Urban Policy Roundtable, February 7.

JONES, E. T. et al. (1975) "Measuring unemployment in the city of St. Louis." University of Missouri (unpublished).
LONG, L. (1980) "Into the countryside and back to the city," in S. Laska and D. Spain (eds.) Back to the City: Issues in Neighborhood Renovation. New York: Pergamon.
——— and D. C. DAHMANN (1980) "The city-suburb income gap: Is it being narrowed by a back-to-the-city movement?" Bureau of the Census, Special Demographic Analysis CDS-80-1. Washington: Government Printing Office.
MOYNIHAN, D. P. (1980) "What will they do for New York?" New York Times Magazine (27 January).
NATHAN, R. P. and C. F. ADAMS (1976) "Understanding central city hardship." Pol. Sci. Q. 91, 1: 47-62.
NATHAN, R. P. and J. W. FOSSETT (1979) "Urban conditions—the future of the federal role," pp. 30-41 in Proceedings of the 71st Annual Convention of the National Tax, Association-Tax Institute of America, Colombus, Ohio.
NATHAN, R. P. et al. (1979) Monitoring the Public Service Employment Program: The Second Round. Washington, DC: National Commission on Manpower Policy.
NELSON, K. (1978) "Movements of blacks and whites between central cities and suburbs in 11 metropolitan areas, 1955-75," in Office of Economic Affairs, Office of Policy Research, HUD, Annual Housing Survey Working Papers: Report # 2. Washington: Government Printing Office.
PETERSON, G. E. (1978) "Fiscally distressed cities: what is happening to them?" Hearings before the Subcommittee on the City, U.S. House Committee on Banking Finance and Urban Affairs, 25 July. Washington: Government Printing Office.
SHALALA, D. (1979) Testimony Before the Subcommittee on Revenue Sharing, U.S. Senate Finance Committee, March 13, 1979. Washington, D.C.
SCHILL, M. H. (forthcoming) "Displacement introduced by urban reinvestment: a research strategy." Policy Paper, Princeton University Urban and Regional Research Center.
SMALL, K. A. (1980) "Energy scarcity and urban development patterns." Princeton University Department of Economics.
STERNLEIB, G. and J. HUGHES (1979) "Back to the central city." Traffic Q. 33, 4: 617-636.
U.S. Department of Labor (1980) Interim Report of the Secretary of Labor on the Recommendations of the National Commission on Employment and Unemployment Statistics. Washington: Government Printing Office.
U.S. Presidential Commission on Employment and Unemployment Statistics (1979) Counting the Labor Force. Washington, DC: Government Printing Office.
VARAIYA, P. and M. WISEMAN (1978) "The age of cities, the employment effects of business cycles, and public service employment," in M. Wiseman (ed.) Studies in Public Service Employment. Berkeley, CA: Institute of Industrial Relations.
VITULLO-MARTIN, J. (1979) "The real sore spot in New York's economy." Fortune (November): 92-105.
U.S. News and World Report (1980) "Comeback for cities, woos for suburbs." March.
VON ECKHARDT, T. D. (1979) "New mood downtown." Transaction (September-October).
Western Resource Area Manpower Consortium (1977) "Surveying unemployment in Cleveland." November.

4

The Fiscal Outlook for Growing Cities

BERNARD L. WEINSTEIN
*Southern Growth Policies Board and
University of Texas at Dallas*
ROBERT J. CLARK
Southern Growth Policies Board

☐ IN DESCRIBING THE FISCAL SITUATION OF CITIES, a presumption is often made that declining populations and economic bases necessarily result in fiscal distress, and that cities experiencing demographic and economic improvements are necessarily better off financially. This is an obvious oversimplification.

The problems of cities are not closely enough tied to readily measureable indicators of change, such as population shifts, to make this an efficient method of monitoring or predicting fiscal distress. Though large decreases in population and economic activity within a jurisdiction are likely to have a negative effect, the fiscal situation of a particular city will also reflect other factors not directly related to growth or regional disparities in growth. Furthermore, growth alone is no panacea for urban fiscal problems. The heterogeneity among growing cities and cities in growing regions rules out such facile distinctions.

Recent emphasis on targeting federal funds has resulted in a number of indexes and distribution formulae to measure the relative need of cities. Due to the complexity of problems faced by cities, however, the use of such indexes may lead to oversimplification in identifying distressed cities and in measuring the level of need. The broad range of urban fiscal, economic,

RECENT DEMOGRAPHIC TRENDS IN AMERICA'S LARGE CITIES

Among the nation's 28 largest cities, 15 lost and 12 gained population between 1970 and 1976. This is a continuation of the pattern of the 1960s. But, the rate of population decline in many older cities was somewhat lower in the period 1973-1976 than in the previous three years, while cities that grew rapidly in the 1960s and early 1970s showed somewhat slower growth rates between 1973 and 1976.

Despite the moderation of population growth rate disparities in large cities, the rate of central city out-migration actually increased between 1975 and 1978 (see Table 4.1). Since 1970, 9 million more people have left central cities than have moved in, a pattern which holds for the South and West as well as the northeastern and North Central regions. In fact, during the period 1975-1978, cities in the South recorded a higher out-migration rate than cities in any other region (see Table 4.2). While the South as a region gained over 1 million persons as a result of migration from northern states between 1975 and 1978, suburban and nonmetropolitan areas were the principal recipients of this migration.

IS A REGIONAL TAXONOMY MEANINGFUL?

Recent demographic and economic trends have favored the Sunbelt over the Frostbelt region; thus, southern and western cities, in the aggregate, have been more robust economically than cities in the Northeast and Midwest. These developments have fostered the notion that urban distress is a regional problem avoided by most growing cities in the South and West. A strong argument can be made for eschewing a regional taxonomy.

Regional aggregation of data on urban conditions may mask the economic and fiscal circumstances of many cities within the region. Population decline, often cited as the precipitator of

TABLE 4.1 Central City Migration, 1960-1978 (in thousands)

	1960-1970	1970-1975[a]	1975-1977[a]	1975-1978[b]
Intra-National[a]				
Central Cities[c]	-3,449	-7,018	-3,321	-4,628
Suburbs	8,756	5,423	2,718	3,527
Total Metropolitan Area	5,307	-1,595	- 603	-1,101
Annual Migration Rate from Central Cities	- 345	-1,403	-1,661	-1,543
Movers from Aborad	N/A	3,604	2,010	2,697
Central City		n/a	786	1,066
Suburban		n/a	792	1,105
Outside States		n/a	732	526
Annual Movers from Abroad		285	393	355

SOURCE: U.S. Bureau of the Census (1975, 1978).

[a]Population 5 years old and over. Central city migration for the 3 years and older age group is about 7,215.
[b]Population 8 years old and over.
[c]Data relate to city boundaries as of 1970.

TABLE 4.2 Migration from Central Cities by Region, 1975-1978 (in thousands)

Region	Intranational	From Abroad	Total	Total Population	Migration as Percentage of Base Population
Northeast	-763	249	-514	14,818	-3.5%
North Central	-1,231	182	-1,049	14,092	-7.4
South	-1,705	285	-1,420	16,600	-8.6
West	-975	377	-598	11,536	-5.2
TOTAL	-4,674	1,093	-3,581	57,046	-6.3

SOURCE: U.S. Bureau of Census (1978).

fiscal distress, is, in many instances, uncorrelated with actual fiscal problems. Another reason for avoiding a regional city classification scheme is that today's liabilities may be tomorrow's assets, and vice versa. For instance, higher energy costs may work in favor of high density cities in the North, while low-density cities in the South and West may find their competitive edge eroded as living costs escalate dramatically in the years ahead. Finally, short-term measures of growth and decline may camouflage real limitations on a city's ability to provide infrastructure and services to its citizenry. Again, there is no particular regional dimension to this problem.

Many of the factors used to identify distressed cities—factors promoting regional taxonomies—may bear little relation to the fiscal conditions of those cities. Age of infrastructure, population decline, and restrictive annexation laws are commonly used as indicators of urban distress. The age of cities, economic and population decline, and fixed city boundaries are highly correlated with urban physical decline and an inability to stem its tide. These considerations would support a regional view of urban distress if it were clear that city age indicated infrastructure conditions, that the economic and population decline which has occurred in the northeastern cities is not spreading to other urban centers, and that continued urban expansion by "growing" cities is indeed possible and beneficial.

Because of their period of development, economic make up, and political constraints, cities in the Northeast and upper Midwest often exhibit a degree of correlation between regional location and urban physical distress. As the region of earliest development, the northeastern quadrant is naturally saddled with the oldest housing stock and oldest physical infrastructure. Yet, infrastructure needs of cities vary in the type of structures and maintenance of existing stock. Whether infrastructures are of comparable quality must be investigated to determine whether age alone indicates greater need.[1] And much of the infrastructure (e.g., mass transit systems) of older cities does not exist in other cities.

Employment and population declines have been much more pronounced in the North. The northeastern quadrant has traditionally been the most heavily industrialized region and the region that contained a preponderance of the manufacturing jobs in the nation. Recent structural changes in the U.S. economy have resulted in an overall decrease in the proportion of manufacturing jobs in the labor force and the interregional redistribution of many manufacturing jobs. The tax losses, direct and indirect, associated with manufacturing decline have posed a problem for all levels of government in the Northeast. The necessary budgetary adjustments needed to bring governmental spending in line with lower revenues, either in absolute dollar terms or when adjusted for inflation, have been difficult for most localities to accomplish. In the first place, it is hard to reduce or eliminate certain public services expected by the corresponding constituencies. Second, although populations in many physically declining cities are dwindling, the younger, more highly educated groups are the most mobile, leaving larger portions of poor, elderly, and minority citizens who are likely to depend on city services. Still, it is possible that the population losses of northeastern cities during the 1960s and 1970s indicate a nationwide trend. Such earlier expanding cities as Los Angeles, Memphis, and Atlanta lost population during the 1970s, and other large cities in the growing region have escaped losses only through massive annexation.

Northern cities are likely to be enclosed by a ring of established suburbs and to be confronted with restrictive state annexation laws. This has made boundary expansion to capture outwardly moving populations and industries more difficult and has removed a possible policy option for cities attempting to adapt to economic and demographic change.[2] On the other hand, statistical indicators of urban conditions for expanding cities or consolidated city/county governments may reflect annexations rather than better overall conditions. Many expanding or consolidating cities, were they to deannex the most physically and socially distressed regions within their boundaries, would be able to reorganize into two cities, one of which would be distressed enough to qualify for existing, targeted federal programs. This situation has been cited in support of efforts to expand existing programs to include "pockets of poverty" criteria for eligibility to otherwise ineligible cities. Without some recognition of these aggregation distortions, considerations of federal program eligibility may be a disincentive to annexation or city/county consolidation, even by those cities to whom these options are available.

Annexation, as a policy to overcome the problems of urban decline, faces other limits. Most annexations have involved unincorporated areas or small suburban communities in need of rehabilitation, unlikely candidates for significantly improving city fiscal conditions. Many formerly expanding cities are now approaching the limits of surrounding incorporated suburbs, hence the potential for future annexations is much less. Even for those cities not so enclosed, there are limitations to their growth. Beyond some optimal size, expanding municipal boundaries may result in incremental servicing costs exceeding potential gains from tax revenue.

Finally, even if a regional taxonomy is appropriate, the fiscal problems of northern cities may be more temporary than permanent. Former Housing and Urban Development (HUD) Secretary Patricia Harris justified the Carter urban policy on the economic point that it is more efficient to revitalize existing cities for use as centers of commerce, industry, and residence than to build new ones. Cities compete with other cities and surrounding areas to retain businesses, industries, and populations and to attract new inhabitants. Because of local and national economic and demographic trends, as well as the inability of many localities to maintain city services and institutions, the competitive advantage of many cities in attracting and retaining corporate and individual residents is threatened. If Secretary Harris' logic holds, many declining cities may be suffering only temporary problems, and the advantages of existing infrastructure and institutions, when combined with federal assistance and a changing environment, may soon present an entirely different set of urban conditions. For example, smaller family size and rising energy costs may be an impetus to locating residences closer to places of work and may also encourage city revitalization as populations and income return from the suburbs. Revitalization is now occuring in many central cities, though usually on a limited scale. If these actions are the beginning of a large-scale "back to the city" movement, cities in the Northeast and Midwest will be able to capitalize on their substantial housing stocks and their more developed infrastructures.

DEFINING URBAN NEED

Cities are also places of residence for many citizens with severe problems (poverty, inadequate housing, high unemployment or underemployment, and low levels of education). To some degree, the two categories of institutional and individual distress afflict all cities. The important distinction between distressed cities and distressed city residents has been largely ignored in developing indicators of urban need, and some would maintain that a misallocation of federal urban assistance has resulted. The causes and manifestations of institutional and population distress are so dissimilar that individual programs are unlikely to resolve or ameliorate both problems simultaneously.

Institutionally distressed cities are burdened with physical infrastructures in need of repair and servicing. The lack of fiscal capability among some cities to maintain, and where necessary to expand, infrastructure has led to a loss of jobs and population, resulting in further erosion of their tax bases and their fiscal vitality. Older cities necessarily developed infrastructures earlier, and those older systems are in greater need of expenditures for repair and maintenance. Older cities are more likely to be surrounded by a developed suburban ring that, with restrictive state laws, has prevented annexation to capture outward migration of industry and population. As a consequence of the recent decline in birth rates, intra- and interregional migration have resulted in declining populations in many large cities.

Socially distressed populations are an old historical problem. Many cities are refuges for many poor, needy citizens. The various problems associated with this type of distress include poverty, inadequate housing, education, employment, and health care. Those cities with large portions of needy citizens face budgetary pressures to provide a wider range of social services, potentially drawing resources away from other functions such as the maintenance of structures.

The degree to which fiscal problems are correlated with other areas of urban distress has been debated by several researchers.

Increased social needs leading to a greater demand for public services, along with economic decline and the associated decrease in tax base, indicate imminent, if not current, fiscal problems. According to a study on urban fiscal stress conducted by Touche, Ross and Company and the First National Bank of Boston (1979), older industrialized cities are most likely to have high tax, debt, and expense ratios and to be financially stressed; but not all such older cities are fiscally distressed, nor is this phenomenon limited to these economically afflicted cities. The study concludes, from this evidence, that "the management and political decision making processes can hold the growth of services in balance with underlying economic resources to maintain financial equilibrium, even under adverse economic, social and structural conditions" (1979: 11). It is further maintained that although economically distressed cities are most likely to be fiscally afflicted, fiscal problems are not inevitable. Social and economic problems can serve as an early warning system of impending fiscal crisis and signal appropriate policy responses to prevent further fiscal deterioration.

Bahl (1979) has attributed a great deal of the fiscal malady to an overdeveloped public sector. Other researchers identify fiscal crisis where too many social needs cannot be addressed because of an insufficient tax base. These conflicting opinions—whether fiscal problems arise from spending too much for city services or from lacking the revenue to spend enough—raise difficult policy questions. The tax potential of a city is difficult to determine and open to dispute. An adequate level of city services is even more difficult to quantify. Clearly, the variations among cities in the level of revenues raised and services provided suggest that this is a policy field deserving much more research and analysis.

IDENTIFYING IMPENDING FISCAL PROBLEMS

Newspaper headlines announcing the impending financial collapse of large U.S. cities, such as New York and Cleveland, have directed public attention to urban fiscal distress. It is, however, a more pervasive problem than a handful of the largest cities being unable to refinance their bonded indebtedness would indicate. Cities of all sizes are finding it increasingly difficult to balance the provision of necessary services with available revenue. Declining populations and tax bases, inflationary price increases, increasing demands for public services, and infrastructure repair or expansion may occur simultaneously and result in fiscal distress for any city. When these problems arise, a city may be unable to balance the provision of necessary services with its ability to pay. Most frequently, this phenomenon manifests itself in financial imbalances, creating further problems for the city in raising money in the capital market. Less obvious, but potentially equally damaging to the life of the city, are unsound expenditure reductions or large tax increases. To balance expenditures with receipts, some cities have reduced outlays for infrastructure maintenance and the provision of services. In the long run, however, these reductions may necessitate greater expenditures to replace structures and facilities that have decayed as a result of this neglect. The lack of adequate services may induce outward migration of population and industry, further damaging the city's revenue producing capacity. Whether fiscal distress is manifested in budgetary imbalances or unsound spending and tax policies, the threat it poses to cities is real.

Growing cities are more likely to have expanding tax bases, but the other factors inducing fiscal problems are no less likely to afflict growing than declining cities. Consumer prices and governmental operating costs are rising fastest in southern and western cities, and, in all likelihood, remaining differentials in prices and costs will have evaporated completely by mid-decade. At the same time, a backlog of demands for infrastructure and improved public services remains unfulfilled in many southern and western cities. Higher-than-average wage increases from growing public sector unionization will burden many city budgets in the growing regions in the years ahead.

FINANCIAL CONDITIONS

Cities faced with expenditures growing more rapidly than revenues may be tempted to finance current consumption with borrowed capital. Although this will relieve temporarily the need to balance outlays and receipts, it merely delays rather than resolves the real crises. By contracting to pay current expenses with expected future income, a city not only imposes costs on future years (requiring even greater adjustments later), but it also limits the flexibility for dealing with future emergencies.

Numerous factors have been cited to explain the expansion of city expenses. The increase in city costs due to inflation is most frequently cited as a major cause of fiscal distress. As city expenses have risen as a result of inflation, tax receipts have not kept pace. Most own-source city revenue comes from property, sales, or income taxes. Graduated income tax revenues typically increase most rapidly, while sales and proportional income taxes generally keep pace with inflation. But property tax revenues, on which most cities rely to a significant degree, are much less likely to keep up with rising municipal outlays. Not only does the time required to re-evaluate property values cause revenues to lag behind expenditure increases, but there is often considerable political opposition to higher property taxes.

The labor intensity of local government also makes it more vulnerable to inflationary pressures. Increasing public employee unionization and the greater incidence of arbitration to settle labor disputes have resulted in upward pressures on public employees wages and benefits, a significant portion of any city budget. Salary and benefit disputes resolved through binding arbitration also lessen elected officials' control over city expenditures.

Cyclical downturns can also adversly effect city governmental finances. A decline in business activity may reduce local tax receipts and at the same time generate increasing demands for municipal services. The fact that local government budgets show an aggregate deficit at the trough of a recession testifies to the sensitivity of city finances to national economic cycles.

The mix of a city's tax sources also affects its revenue elasticity. The efficacy of income or sales taxes in raising revenues that keep pace with increasing costs during inflation is tempered by the vulnerability of these sources to an economic slowdown. Job losses and lowered retail sales accompanying a recession can significantly lower revenues from income and sales taxes. Property taxes, whose stability makes them less-than-ideal revenue sources during inflationary periods, are less reactive to economic downturns (except during significant, long-term depressions where considerable change of ownership takes place and property revaluations are made downward).

LEVELS OF FISCAL DISTRESS

The fiscal crisis of the cities is more pervasive than two or three large cities tottering on the brink of bankruptcy. Broadly, it covers a number of situations where demographic, economic, and political realities have placed cities in less-than-advantageous financial positions. It is more than the simple problem of too few resources available to make all desired purchases of goods and services. City governments exist to allocate scarce resources to their most beneficial purposes, and no action or actions can be developed to create sufficient resources to meet all demands. Rather, fiscal distress is a number of problems placing some cities and their governments in more difficult positions and limiting their financial options. Various researchers have identified three categories of financial distress; relative fiscal disadvantage, fiscal decline, and acute fiscal crises (Barro, 1978).

Fiscal disadvantage describes the situations of cities less able to provide given levels of public services than other, similarly situated cities. It may arise from one or more of the following causes: low tax base, high level of service demands, high costs of services, or a broad range of fiscal responsibilities. Fiscal disadvantage is characteristic of many central cities in relation to their surrounding suburbs. If these disparities become too great, cities with less tenable fiscal situations may become increasingly undesirable as residential or business locations.

Fiscal decline is a relative or absolute deterioration of budgetary options for urban local governments. Relative decline indicates an expansion of fiscal capacity less rapid than in other cities. It may accompany economic expansion, but it usually indicates a loss of comparative attractiveness. Relative decline may eventually lead to absolute losses of people and business to jurisdictions with lower taxes or higher service levels. Absolute decline describes a situation where a given level of taxation will provide fewer services because of a decline in the real tax base, a drop in the real value of intergovernmental transfers, a higher proportion of above average service consumers (e.g., the poor and school-aged children), or an increase in service costs. (Rising city taxes do not necessarily indicate absolute fiscal decline. If the city's tax base is inelastic—not expanding as rapidly as community wealth—tax rates may rise without increasing the absolute level of tax effort.) A city facing absolute fiscal decline must adjust resources and service levels to regain a financial balance. Cities that cannot or will not make these adjustments are likely to be confronted with acute fiscal crises.

Acute fiscal crisis describes true fiscal emergencies where the problems leading to fiscal distress are too severe or adjustments have been too long delayed to be resolved by ordinary budgetary mechanisms. A crisis results in an abrupt disruption of normal revenue and expenditure patterns and a failure to meet contracted debt payments. The distinction between this and other levels of fiscal distress is most obvious in cases such as New York City. For too long, New York delayed readjustment in service level provision to bring it in line with current revenue receipts. Successive city administrations postponed the difficult decisions of either reducing services or imposing higher taxes by financing current operations with borrowed funds. Eventually, lenders became unwilling to either refinance existing debt or finance new current account deficits. Facing this dilemma, the city turned to special state and federal assistance to avert financial collapse. Barro describes the consequences of acute financial crisis as: (a) more severe service disruption than would have been required had adjustments been instituted earlier; (b) shocks to the urban economy from large increases in the number of unemployed when public employment is drastically reduced; (c) disruption of the financial markets; and (d) loss of local autonomy as outside governments are granted extended authority over local decisions for having intervened to stave off forfeiture and fiscal collapse (1978: vi).

Each of the three levels of fiscal distress indicates a progressively worsening situation. Although most analyses of urban fiscal conditions have assigned a low probability to widespread financial collapse, less severe forms of fiscal distress may crop up in many cities.

INDICATORS OF FISCAL DISTRESS

A number of methods of measuring the fiscal health of cities have been developed to assist local efforts in combating urban decline and to direct federal assistance. The simplest of these is the National Income Accounts aggregated budget surpluses (deficits) as an indicator of the overall city financial situation. These NIA figures, which showed an overall deficit during parts of the 1974-1975 recession, have demonstrated a remarkable recovery in recent years. The total NIA surplus reported for the state-local sector in the first quarter of 1978 was $30.2 billion. For a number of reasons, these figures can be misleading. First, the budget balances of cities and states are combined, masking the true condition of cities. Second, pension-fund surpluses are included in the assets of state and local governments, although they are not available for use in governmental operations. Third, operating surpluses are necessary for many jurisdictions because of restrictions against deficit financing and the distinctions in time between revenue collection and the outlay of expenditures. Finally, a healthy fiscal condition for all jurisdictions cannot be inferred from the existence of an aggregate operating surplus.

An analysis of the fiscal conditions of a sample of large U.S. cities was recently conducted by the First National Bank of Boston and Touche Ross and Company (1979). According to this report, the fiscal situation of most cities is good. The study examined 120 cities with populations of less than 1 million. They collected indicators of economic, social, and structural conditions, as well as municipal performance variables, for the cities. Due to data unavailability, 54 of the cities in the sample had to be eliminated. The 66 remaining were grouped into clusters based on the indicators of economic, social, and structural conditions. Municipal performance variables, tax performance, debt performance, and expense performance for each city were compared with measurements of central tendencies for other cities in the same cluster. To qualify as fiscally distressed, a city had to fall more than one standard deviation above the cluster mean for each of three municipal performance variables. Only four cities—Stamford, Boston, Hartford, and Atlanta—fell into this category. Eight additional cities—Denver, Bloomington (Minnesota), Seattle, Worcester, Duluth, Long Beach, Richmond, and Fresno—were outside one standard deviation on two of the three variables.

From this analysis, the report arrived at a number of conclusions and policy recommendations. Older, industrially aged cities, marked by manufacturing employment decline during the 1950s and 1960s, are most likely to have high tax, debt, and expense ratios and to be fiscally stressed. But not all older cities are so afflicted. Many younger, growing cities also indicate incipient fiscal distress. From these observations, the authors concluded that fiscal distress is not an inevitable result of economic maturity. Through sound fiscal management and resourceful decision-making keeping the growth of services in balance with underlying economic resources, financial equilibrium can be maintained. The study further observes that although economic, social, and structural conditions may properly be used as criteria for distributing federal assistance funds, they are insufficient for determining fiscal conditions and appropriate levels of federal fiscal assistance. Financial data must also be considered. Municipal data collection and financial reporting systems are severely criticized for their lack of uniformity and clarity. The procedures of many cities, the study maintains, do not provide sufficient information to allow local governments to take appropriate actions to maintain fiscal solvency.

The Touche Ross-First National Bank of Boston study has received considerable criticism in its methodological approach and in the results. The Office of Policy Development and Research of the U.S. Department of Housing and Urban Development attacked the assumption that city fiscal health can be measured from a set of budgetary variables (1979): "By itself a city's balance sheet does not measure the economic well-being of the jurisdiction nor does it necessarily indicate the future financial circumstances of the jurisdiction. ... The goal of the city ... is to provide service to people—a goal which cannot be measured by a city's accounting record alone."

The study also was criticized for its sample selection (cities over 1 million were excluded, although some of the most severe fiscal problems exist for these cities) and for defining fiscal distress in relation to the financial situation of similar cities in a "cluster." By assuming that variation beyond the cluster mean by more than one standard deviation indicates fiscal stress, the researchers would be unable to detect fiscal stress should all cities in the cluster be so afflicted. The study only compares central cities to one another while ignoring the relative position of the sample cities to their surrounding suburbs. Finally, HUD criticizes the report for using 1975 data to analyze fiscal stress in 1979.

The criticisms of the Touche Ross report have obvious merits; nonetheless, the report does provide some perspectives for gauging imminent fiscal dangers. First, the fiscal situations of cities are not necessarily determined by the economic and social conditions of city residents. Second, local fiscal health is strongly affected by the response of municipal governments to changing economic circumstances. If cities carefully monitor their changing economic circumstances, and make the necessary expenditure and tax adjustments, fiscal solvency can be maintained.

Evidence of fiscal distress occurring in growing as well as declining cities points out the weakness of the hypothesis that decline leads to financial problems. The relatively more secure fiscal position of growing cities might be explained, however, if both fiscal health and growth are assumed to be the result of common factors and practices on the part of states and cities. Among the plausible explanations for fiscal health among cities in the growth regions are fewer governmental services, larger state roles in financing, a greater reliance on user fees rather than taxes to finance some city services, and a perception of a better business climate.

Breckenfeld (1977) described three cities' actions to reduce governmental services and expenditures to bring them in line with smaller populations and tax bases. In each—Pittsburgh, Baltimore, and Wilmington (DE)—Breckenfeld ascribes the success in reducing expenditures and dealing with city shrinkage to the political leadership of the mayors. The actions of Pittsburgh's Mayor Flaherty are exemplary. To bring city expenditures in line with lower revenues, Mayor Flaherty cut city payrolls (mainly through attrition) and confronted city unions with demands to reverse union-induced activities the mayor viewed excessive. The activities of Baltimore's Mayor Schaefer and Wilmington's Mayor Maloney in combining city services, reducing city payrolls, and developing incentives for city revitalization are also cited as potential models for other cities facing smaller futures.

SUMMARY AND CONCLUSION

There is an obvious link between urban population change and fiscal conditions. The highly visible incidence of population decline and fiscal instability in such major cities as New York, Cleveland, and Newark has been taken as prima facie evidence that population decline leads to fiscal decline. Dallas and Houston are cited as examples of the nexus between growth and fiscal health. While we do not argue that there is no relation between changes in population size and fiscal condition, we do maintain that there are other explanations of fiscal distress.

Most simply put, fiscal conditions are a function of the revenue-raising potential of a city balanced against the costs of providing a chosen level of goods and services to its citizenry. Population decline and industry loss, resulting in fewer jobs and further population decline, will inevitably lower revenue. The level of revenue reduction should, after a period of adjustment, be at least partially matched by reduced expenditure requirements. But the greater mobility of younger, better educated groups does leave a more service-demanding populace of the elderly, the poor, and the less well-educated. Yet bankruptcy may not be the inevitable result. Beyond the examples of New York and Cleveland, large cities which have had large population losses and have tottered on the brink of financial collapse, stand other examples of cities that, while suffering similar population declines, have avoided such potentially ruinous financial debacles. To adapt to declining revenues resulting from lower populations, Pittsburgh and Baltimore have reexamined the level and type of services offered, reducing and cutting back to bring expenditures in line with receipts. Minneapolis, Saint Paul, and their surrounding area have devised a tax base sharing scheme that co-opts the suburbanization of population and industry, much as the more formal city-county consolidations in Indianapolis, Nashville, and Jacksonville have done.

Beyond the effect of population decline on fiscal problems, the reverse must also be considered. Cities that have responded to fiscal imbalance by continually raising taxes are also frequently losing population. The high tax levels, where raising revenue corrects budgetary problems, have damaged the competitive position of many cities to attract and hold industry and population in relation to their suburbs and cities in other regions. Thus, the causal link between population and fiscal health, and the extent it can explain fiscal conditions, is not unidirectional.

We must also recognize that there are important differences between growing/fiscally healthy cities and shrinking/fiscally troubled ones other than their ability to attract and keep populations. One difference is the more liberal state annexation laws mentioned above. Potentially more important is the general tendency of most large, declining cities to offer more and more expensive services than do most growing cities. There is no method to ascertain the "correct" level of local government services. Indeed, regional distinctions in taste and tradition may make such comparisons meaningless. Nevertheless, the examples of cities with growing tax bases and limited spending habits, as well as declining cities who have improved their fiscal position by reducing service levels, do point out the influence not only of revenue raising capability but also of the incentive to spend on the fiscal health of cities. Another influence on urban fiscal conditions is the method of financing urban services. The more common use of service charges among growing cities relieves the tax burden of subsidizing a specific governmental service consumed by only a portion of the citizenry.

In projecting future urban trends, there are reasons to assume that older, declining cities may be in an improved position, while some of the cities that have grown over the recent past may face new problems.

If it turns out that federal assistance plans have been efficacious in dealing with urban problems, the efficiency with which these funds have been targeted to cities with declining populations bodes well for their future. Federal programs, such as the Community Development Block Grant and the Urban Development Actions Grant programs, are aimed at revitalizing declining cities and assisting in reconstructing tax bases. Urban homesteaders, relatively affluent individuals recapturing residential neighborhoods in many major cities, may also contribute to improved fiscal conditions. Rising transportation costs and fuel prices are adding to the impetus to return to close-in residential locations. The type of housing stock most frequently renovated is in much larger supply in many of those cities currently suffering the largest declines in population. If this trend of neighborhood revitalization continues, for fiscally troubled cities a significant improvement in the local tax bases is predictable.

For the central cities of the growing metropolitan areas of the South and West, the 1980s will probably witness slow or negative population growth and increased fiscal pressures. Birmingham, Atlanta, Memphis, New Orleans, Denver, and Los Angeles all lost population during the 1970s and this trend is likely to continue. Central city manufacturing employment has also dropped in those cities, with most of the employment gains accruing to suburban areas. If the suburbs of growing metropolitan areas continue to attract the lion's share of population as well as commercial and industrial development, increased local tax receipts from these sources may result in noticeable fiscal disparities between central cities and suburbs on the revenue and expenditure sides. Sunbelt cities may find themselves afflicted by the economic and fiscal erosion plaguing the older industrial cities of the Northeast and Midwest.

Despite slower population growth, demands for greater expenditures by local governments in the South and West will not abate. To a large extent, the local public sector in the growing regions remains "underdeveloped." That is, local governments in the South and West typically provide a narrower range (and perhaps lower quality) of public and social services than is the case elsewhere. But ten years of unparalleled growth have resulted in a backlog of demands for new public infrastructure and better public services. Higher per-capita incomes will also generate pressures for increased local fiscal activity, as will growing unionization of public employees across the Sunbelt.

The costs of living and working in Sunbelt cities have been escalating rapidly over the past decade, and by the end of the 1980s it seems probable that cost differentials between northern and southern cities will be virtually nonexistent. There is already strong evidence to support this prediction. Since 1972, the consumer price index (CPI) for southern cities has increased at a faster rate than for any other region of the nation. By contrast, the Northeast has posted the slowest price gains among major regions. Houston has recorded the sharpest price increases of any major city in the United States since 1972, while New York City has posted the smallest. Production costs are also rising faster in the Sunbelt than in other regions. Since 1975, average hourly earnings in manufacturing have grown at a faster rate in the South and West than in the Northeast and North Central regions. The fact that consumer prices and production costs in the South are catching up to those in the North is an inevitable, if unfortunate, consequence of the region's rapid growth combined with higher energy costs. And this convergence will probably slow down the inward migration of people and industry to southern and western metropolitan areas in the decade ahead.

NOTES

1. It is unclear whether there is a significant correlation between age and quality. The Department of Housing and Urban Development justifies the use of the housing age variable as an indicator of overall infrastructure age, which will more likely have higher maintenance costs.

2. The degree to which annexation actually improves the overall position of cities and their residents is not altogether clear. Through annexation of mobile populations and industries, cities should be able to better maintain their fiscal positions.

REFERENCES

BAHL, R. (1979) "The New York State economy: 1960-1978 and the outlook," Occasional Paper 37, The Maxwell School, Metropolitan Studies Program. Syracuse, NY: Syracuse University.

BARRO, S. (1978) The Urban Impacts of Federal Policy: Volume 3, Fiscal Conditions. Santa Monica, CA: Rand.

BRECKENFELD, G. (1977) "It's up to the cities to save themselves." Fortune (March).

Touche Ross & Co. and National Bank of Boston (1979) Urban Fiscal Stress: A Comparative Analysis of 66 U.S. Cities. New York: Touche Ross.

U.S. Bureau of the Census (1978) "Geographic mobility: March 1975 to March 1978." Current Population Reports Series P-20, 331.

——— (1975) "Mobility of the population of the U.S.: March 1970 to March 1975." Current Population Reports Series P-20, 285.

U.S. Department of Housing and Urban Development, Office of Policy Development and Research (1979) Urban Fiscal Crisis: Fantasy or Fact, 26 March. Washington, DC: Government Printing Office.

5

The Tax and Expenditure Limitation Movement

DEBORAH MATZ
Joint Economic Committee

☐ JUNE 6, 1978, THE DAY PROPOSITION 13 was overwhelmingly passed by the California electorate, was the birth of the "tax revolt" movement in the United States. Economist Paul Samuelson acclaimed it the "most important political-economic event of 1978, perhaps even the 1970's."[1] After its passage, numerous and varied measures to limit taxing and spending at all levels of government were considered by the electorate in a majority of states.

While Proposition 13 type of revenue restrictions were passed in three of four states in 1978, five states approved state or local government expenditure limitations, including Tennessee's state spending limit adopted in March, 1978, prior to the passage of Proposition 13. By November 1979, state revenue or expenditure limitations were imposed in eight more states (See Appendices A and B). This chapter will examine the efforts to limit revenues and expenditures at the state and local levels, the types of limitations, their rationale, and their likely effect on state and local finances.[2]

The various state and local government tax and spending limitations tend to be considered as a group. But there is a good deal of difference between and among each category: There are limitations on expenditures and revenues; some are aimed at reducing current levels of revenues or expenditures; others concentrate on slowing their growth; and some limit only local governments or states, and others, limit both.

Tax limitations on the state and local levels restrict the jurisdiction's ability to raise own-source revenue. On the local level, property tax collections typically are restricted. This analysis will deal with only the most recent and severe type of local revenue limit inspired by Proposition 13. It will not examine tax rate changes, levy limits, or other restrictions on local revenue flows.

In California, Idaho, and Nevada, where severe local tax restriction measures were approved by the electorate, the limit on taxes collected is achieved by rolling back property assessments, specifying future allowable percentage increases in assessments, and requiring approval of the voters or legislature to approve future increases or new taxes. State revenue limits restrict growth rather than reduce revenues. Three states have passed limits on state revenues. In Michigan, Washington, and Louisiana, state revenue increases are tied to the growth in state personal income.

Spending limitations, the more popular of the two limitations, restrict state or local government expenditures in a given year by tying annual expenditure increases to some measure of growth in the State's economy, most frequently state personal income or state gross product.

Exceptions include Nevada (where state spending is tied to the level of the preceding year's budget adjusted for inflation and population), South Carolina (prohibited from spending more than 95% of its current year's revenues), and California, (where spending is limited to the inflation rate plus population growth).

The diversity of the tax and expenditure limitation measures makes it difficult to draw anything but broad conclusions about the causes and effects of the limitation movement. For this analysis, all expenditure limits are considered as a group, as are all revenue limits, despite technical differences in implementation. Separate analyses for state and local initiatives are provided, as feasible.

Investigating the causes and effects of the tax and spending limitations raises several immediate questions. Why does a jurisdiction select revenue limitations instead of expenditure limitations and vice versa? Is there a differential effect of imposing one instead of the other? Why are the limits imposed on local governments in some instances, state governments in others, and both in yet other cases? Is the motive similar in all situations? What is the anticipated and actual effect? The limitation movement is so new and the motive and potential effects so complex that the literature offers little consensus in response to these questions.

TAX BURDENS

The Washington State legislature, in formulating a limitation policy, was advised that if spending is the source of concern, expenditures should be limited; if rapidly increasing taxes are troubling, a revenue lid should be used (Rice, 1978). From this, one might surmise that tax and expenditure lids have been adopted by states particularly afflicted by high or rapidly increasing taxes or expenditures. This conclusion, however, is not consistent with the data in all cases.

Of the six states enacting revenue limits in 1978 or 1979, only California's property tax per $1,000 of personal income exceeded the U.S. average in 1967; Michigan joined California in exceeding the U.S. average in 1977 (see Table 5.1).[3] Of the four States passing revenue limits in 1978, Idaho, Nevada, and Michigan were below the average property tax burden for the nation in 1967. The relative property tax burden in both Idaho and Nevada dropped considerably in 1977, from 98% to 80% and from 98% to 89%. Even compared to the other states in their region, the property tax burdens in Idaho and Nevada were well below the average in both 1967 and 1977. Similarly, the property tax burden per $1,000 of state personal income was below the national average in all the states enacting revenue limits in 1979, and in two of the three states, the property tax burden in relation to the national average decreased in 1967-1977.

Of the states adopting revenue limits in 1978 or 1979, per-capita property tax revenue exceeded the national average in 1967 in California, Michigan, and Nevada. But in 1977, only California and Michigan's outpaced the national average (see Table 5.2). Even in California, the average annual rate of

TABLE 5.1 State and Local Property Tax per $1000 of State Personal Income, 1967 and 1977

	1967		1977	
	Amount	Percentage of U.S. Average	Amount	Percentage of U.S. Average
States enacting revenue limits in 1978				
Local:				
California	62	138	65	141
Idaho	44	98	37	80
Nevada	44	98	41	89
State:				
Michigan	43	96	49	107
States enacting revenue limits in 1979				
State:				
Louisiana	24	53	19	41
Washington	35	78	38	83
United States:	45	100	46	100
Southeast	27	60	25	54
Rocky Mountain	56	124	50	109
Far West	48	107	50	109
Great Lakes	45	100	44	96

SOURCE: ACIR (1978-1979: vol. M-115, Table 41)

increase from 1967 to 1977 did not exceed the national average. Only Michigan and Washington's were more rapid than the national average.

When considering total tax revenue in relation to state personal income (in five of the six states adopting revenue limits in 1978 or 1979), tax revenue as a percentage of personal income exceeded the national average in 1965 (see Tables 5.3 and 5.4). By 1977, only three of the states—California, Nevada, and Michigan—exceeded the U.S. average. In fact, in every state but California, tax effort decreased relative to the national average between 1965 and 1977. Again, California was the only state of the six where the annual average increase in tax revenue as a percentage of personal income exceeded the national average.

TABLE 5.2 Per-Capita State Local Property Tax Collections

	Amount 1967	1977	Average Annual Rate of Increase 1967-1977
States enacting revenue limits in 1978			
Local:			
California	209	458	8.2%
Idaho	108	205	6.6
Nevada	150	286	6.7
State:			
Michigan	139	332	9.1
States enacting revenue limits in 1979			
State:			
Louisiana	54	99	6.2
Washington	111	255	8.7
United States:	132	289	8.2
Southeast	60	134	8.4
Rocky Mountain	150	295	7.0
Far West	155	338	8.1
Great Lakes	141	288	7.4

SOURCE: ACIR (1978-1979: vol. M-115, Table 42)

LEVEL AND GROWTH OF EXPENDITURES

The pattern is similarly unclear in those states adopting expenditure limitations in 1978 and 1979. Of the five states that adopted limits in 1978—Arizona, Hawaii, and Michigan—state expenditures for selected state and local functions from own-source revenue as a percentage of state personal income exceeded the national average in the periods 1974-1975 and 1976-1977, while Tennessee and Texas were below the national average in both years (see Table 5.5). But, when considering the average annual percentage change in expenditures for selected state and local functions from fiscal year 1975 to fiscal year 1977, only Hawaii and Texas experienced a more rapid change in spending than the aggregate change in spending by all states.

TABLE 5.3 State and Local Tax Revenue in Relation to
State Personal Income, 1965 and 1977: Tax
Revenue as Percentage of Personal Income

	1965	1977	Annual Average Percentage Change
States enacting revenue limits in 1978			
Local:			
California	11.98	15.49	2.2%
Idaho	12.14	11.70	-0.3
Nevada	10.69	12.93	1.6
State:			
Michigan	10.67	13.04	1.7
States enacting revenue limits in 1979			
State:			
Louisiana	12.05	12.01	0.0
Washington	11.18	12.23	0.8
United States:	10.45	12.80	1.7
Southeast	10.04	10.91	0.7
Rocky Mountain	11.61	12.99	0.9
Far West	11.79	14.84	1.9
Great Lakes	9.73	11.72	1.6

SOURCE: ACIR (1978-1979: vol. M-115, Table 24)

Of the five states adopting expenditure limitations in 1979, two were above and three were below the national average expenditure-income ratio in 1974-1975 and 1976-1977 (see Table 5.6). Three of the five also experienced a more rapid average annual growth in expenditures than the average annual growth experienced by all states. Based on these data, the evidence is still inconclusive that limitations are a response to high expenditures or rapid expenditure growth or to high property tax burdens or rapidly increasing tax revenues.

OTHER FACTORS

These results add to the existing confusion over the causes of the overwhelming support of California's Proposition 13 and other tax and spending limits. Because limits were adopted in

TABLE 5.4 State and Local Tax Revenue in Relation to State Personal Income, 1965 and 1977: State Percentage Related to U.S. Average

	1965	1977
States enacting revenue limits in 1978		
Local:		
California	114.6	121.0
Idaho	116.2	91.4
Nevada	102.3	101.0
State:		
Michigan	102.1	101.9
States enacting revenue limits in 1979		
State:		
Louisiana	115.3	93.8
Washington	107.0	95.5
United States:		
Southeast	96.1	85.2
Rocky Mountain	111.1	101.5
Far West	112.8	115.9
Great Lakes	93.1	91.6

SOURCE: ACIR (1978-1979: vol. M-115, Table 42)

states not experiencing high or rapidly escalating taxes or expenditures, the movement is not necessarily an attempt to address these problems.

Since the recent limitation movement began, there has been a great deal of debate but little agreement on the causes of the movement's widespread popularity. The editors of the proceedings of a recent conference on the causes and consequences of tax and expenditure limitations concluded that the meeting produced little consensus (Shapiro et al., 1979).

Some see the movement as a means of reducing taxes but not services (Shapiro et al., 1979: 9). A study conducted in California prior to the vote on Proposition 13 supports the view that the voters did not generally support across-the-board spending cuts (Citrin, 1979). In fact, they tended to favor *increased* expenditures for police, fire, mental health programs, and education. In only three of fifteen areas did citizens desire cuts: welfare, public housing, and the government's own administrative services.

TABLE 5.5 State Expenditures for Selected State and Local Functions from Own-Source Revenue, 1974-1975, 1976-1977 (states enacting state spending limits in 1978)

	1974-1975		1976-1977		Average Annual Change
	Amount	Amount as Percentage of State Personal Income in 1974 Related to U.S. Average	Amount	Amount as Percentage of State Personal Income in 1976 Related to U.S. Average	
Arizona	722	114	817	109	6.4%
Hawaii	438	149	520	147	9.0
Michigan	3,446	112	3,851	110	5.7
Tennessee	998	92	1,160	90	7.8
Texas	2,630	77	3,400	77	13.7
United States	66,000	100	78,094	100	8.8

SOURCES: ACIR (1976-1977: vol. M-113, Table 8; 1978-1979, Vol. M-115, Table 11).
NOTE: Dollars are in millions.

TABLE 5.6 State Expenditures for Selected State and Local Functions from Own-Source Revenue, 1974-1975, 1976-1977 (states enacting state spending limits in 1979)

| | | 1974-1975 | | 1976-1977 | |
	Amount	Amount as Percentage of State Personal Income in 1974 Related to U.S. Average	Amount	Amount as Percentage of State Personal Income in 1976 Related to U.S. Average	Average Annual Change
California	6,874	95	8,091	92	8.5%
Nevada	151	76	179	72	8.9
Oregon	572	83	701	84	10.7
South Carolina	841	122	937	112	5.6
Utah	312	103	408	109	14.4
United States	66,000	100	78,094	100	8.8

SOURCE: ACIR (1976-1977: vol. M-113, Table 8; 1978-1979: vol. M-115, Table 11)
NOTE: Dollars are in millions.

A Michigan survey concluded much the same (Courant et al., 1979). More than 50% of the respondents did not favor any reduction in state and local spending or taxes. Of the six programs considered, welfare was the only program for which the number of people desiring reduced expenditures exceeded those desiring increases. Those who expressed a desire for cutbacks sought only minimal cuts, but, with the exception of welfare, more respondents favored current or increased levels of spending.

The study also considered the perception about tax limitations and found that the strongest support for Michigan's tax limitation amendment came from people who felt it would increase governmental efficiency or voter control of government. Thus, it is not as clear that reduced governmental expenditures and tax reform are widespread goals of the limitations movement. For many, it an attempt to achieve bureaucratic reform—reducing the size of the public sector, reducing its inefficiency, and equalizing the tax burden.

Simms concurs (1978: 325). She does not ascribe the movement to a desire to reform the tax system but speculates that the property tax is a symbol of something else—possibly displeasure over the size of the public sector, malcontent over the composition of the public sector, a desire to shift the tax burden to the State, or a concern for inequities in administrating the tax system.

But, there are those who believe that, regardless of other factors, the motive is the limitation of revenues and expenditures (Suyderhoud, 1978; Ladd, 1979; Advisory Commission on Intergovernmental Relations [ACIR], 1977a). Those who support this theory argue that the rapid increase in state and local expenditures in the 1960s and early 1970s resulted in governmental spending getting out of control. At the same time, increases in property and income taxes have angered taxpayers who see their taxes rising more rapidly than their incomes.

In light of these findings, the current limitation movement is a response to a network of stimuli, rather than to one particular occurrence. Among the possible other causes is inflation, causing property values to rise. Prompt property reassessments will, therefore, increase taxes without an increase in tax rates. Similarly, income taxes increase as inflation pushes taxpayers into higher tax brackets. Thus, even if the income tax is a flat rate, taxes are increased as income, due to inflation, increases. Nine of the states limiting spending or revenues in 1978 or 1979 have income taxes, and in all but one instance, the tax is progressive—although it is more progressive in some states than in others. Residents in these states, therefore, are pushed into higher federal and state brackets by inflation, thereby losing an increasing proportion of their income to taxes as their earnings increase. As a result, many citizens, believing they are paying too much for too little, look to the limitation movement to reverse this trend.

In recent years, there also appears to be increasing frustration with all levels of government—the red tape, regulations, and the apparent waste and inefficiency. Moreover, citizens seem to be dissatisfied with the decisions on taxing and spending emanating from the representative political process.

Those who believe that the dissatisfaction with government and the existing decision-making process is responsible for the current trend would probably see the goal as a major change in managing government—transfering the tax and spending decisions from the elected officials to the people. Whether or not this is the goal, it is perhaps the most significant and potentially serious consequence of the movement. It not only indicates a widespread lack of faith in the existing representative system, but it severely limits the ability of elected officials to govern.

More likely than not, no one of these factors is responsible for the strength of the limitation movement. Whereas the high residential tax burden in California may have been decisive in passing Proposition 13, and the rapid increase in state expenditures may have provoked support for a spending lid in Texas, these circumstances are not universal and do not account for the popularity of the movement in a multitude of states.

HISTORICAL PERSPECTIVE

Despite the fact that the movement to limit taxes and spending has produced a great deal of media attention and scholarly study, fiscal limitation is not new. Perhaps the new attention and concern can be explained by the type of limits and the more dramatic manner they have been adopted. Generally, efforts have involved statutory initiatives approved by the state legislature. This method is still popular, but a number of recent limitations have been amendments to the state constitutions passed by general referendums. Unlike statutory limits, these are adopted by the voters and are not easily changed.

Further, pre-1970 efforts typically involved local property tax rate limits. The first rate limits adopted at the end of the nineteenth century limited the growth of local government expenditures (ACIR, 1977). Another occurred in the 1930s when pressure was exerted to reduce property tax rates and mitigate the increasing rate of tax delinquencies. Seven states responded by revising their tax rate limits—the traditional method by which states restrict local government financing. Prior to 1970, all but 10 States had enacted such limits (Shannon, 1979).

In addition to limits on property taxes, states often restrict local governments from imposing other taxes. As of 1976, local governments in 18 States were not authorized to adopt sales taxes, and in 39 States could not impose income taxes (ACIR, 1977). The 1970s, however, has been a decade of innovative fiscal controls. Some states have indexed income taxes to moderate tax increases caused by inflation. Many states limited local revenues through property tax circuit breakers intended to prevent an excessive tax burden on certain property owners— (generally the aged and lower income), and new classification laws designed to relieve property tax in select, economic activities. Further, limits on property tax levies, which restrict the growth of total property tax revenues to a specified annual increase, increased dramatically in the 1970s. While, prior to 1970, only Arizona, Colorado, and Oregon had adopted tax levy limits, by September 1979, 20 states had done so (Shannon, 1979).

Full disclosure or truth-in-taxation laws are yet another form of property tax revenue restrictions and also a product of the 1970s. These laws, more restrictive than tax rate or levy limits, are the precursor of the recent, more severe, tax and spending limitations. Full disclosure laws require the establishment of a property tax rate providing property tax revenues equal to the previous year's, or, in some instances, a specified limit above that amount. Often, property tax rate roll-backs are required to offset additional revenues resulting from increased assessments.

In all instances, local elected officials must announce their intention to increase rates, and the locality must hold public hearings before taking action. Florida was the first to enact a full disclosure law (1971). By 1979, such laws had been enacted in nine other states: Arizona, Hawaii, Kentucky, Maryland, Montana, Rhode Island, Tennessee, Texas, and Virginia. Five of these states also enacted strict state revenue or spending limits in 1978 or 1979: Arizona, Hawaii, Kentucky, Tennessee, and Texas.

Although spending limitations did not receive much public attention before 1978, they are by no means new. The first local spending limit was adopted in 1921 by Arizona. The Arizona legislature imposed a 10% annual cap on growth for city, county, and town expenditures. It was not until 1976 that another state imposed restrictions. The New Jersey measure, unlike many of the recent limits approved by the electorate, was also passed by the state legislature. It limited local expenditures to 5% over the previous year, and restricted state expenditures to the annual increase in state per-capita personal income. Following the New Jersey precedent, the Colorado legislature enacted a state spending limitation in 1977 by voting to place a 7% cap on the annual increase in state expenditures.

The movement for fiscal limitation, then, is a continuum, begun to moderately restrict taxes and remaining relatively unchanged for almost a century. Recently, particularly in jurisdictions with frequent reassessments, property taxes have increased substantially, rendering rate limits ineffective in containing property tax bills. Thus, the rate limits gave way to less moderate measures—circuit breakers, levy limits, indexation, and full disclosure laws—limiting certain tax revenues. The recent, more severe, measures actually reducing tax revenues in some instances and limiting total revenues in others, though extreme, have not materialized overnight. They have evolved from a lineage of fiscal controls. Indeed, even the expenditure limits, the most recent and innovative control, is an extension of state constitutional requirements for balanced budgets and other existing initiatives limiting borrowing, interest rate ceilings on bonds, and access to (and the extent of) usage of certain taxes.

EFFECTS

As there exists a range of opinions on the causes of the tax and expenditure limitation movement, so there is diversity of opinion on its likely effects. Some believe limitations will curb expenditures, increase efficiency and accountability, and stimulate economic development. Others foresee reduced public services and work-forces and a proliferation of special districts and user charges.

The ACIR (1977b: 21) found that, while tax limits do not reduce the dependency of local governments on property taxes as a component of own-source revenue in states with limits (rather than those without limits), they restrain per-capita local expenditures. The Commission suggested that local governments restrained by tax limits *ceteris paribus* spend an average of 6-8% per capita less. Ladd (1979), however, concluded that local tax and expenditure controls did not limit costs, particularly when the cause of rising costs is the lagging productivity of the service sector, the lack of market competition, and the growth in direct service requirements per unit of output associated with deteriorating environmental factors. Ladd also found that the benefits of local controls were slight, while service level distortions that they create may impose significant costs. Finally, insofar as implementing taxing controls to reduce the property tax burden, she found that under certain conditions, temporary controls were justified when combined with additional state aid.

Thus, the two major studies arrived at different conclusions. Because they were conducted before the current, more severe, limitation movement, it might be instructive to consider the results of California's Proposition 13, both in its actual impact and to illustrate the difficulty in anticipating potential effects.

Prior to its passage, proponents argued that lower property taxes would result in new investment, generate higher profits and personal income, create more private employment, and slow the growth of the public sector (D. J. Levin, 1979). Opponents, on the other hand, painted pictures of drastically reduced expenditures for public services and education and massive layoffs of public employees. Neither of these predictions was entirely true. From the second quarter of 1978 to the second quarter of 1979, California experienced an increase in personal income slightly above that of the nation, but not as much above the national average as in 1977; the state increase in retail sales exceeded the U.S. average by 5.5%, while in the past two years it had paralleled the national average; and private employment increased 5% in California compared to 3.5% in the U.S. (Levin, 1979: 14). This is less than the difference in 1978 and slightly more than in 1977. While the improvement in California's economy, predicted by proponents of Proposition 13, did not materialize; neither did the drastic reductions in the public sector. Levin estimates that local government employment from mid-1978 to mid-1979 declined by about 60,000 with the period from January 1979 to mid-1979 reflecting some recovery from the 100,000 decline in employment that occurred up to that point.

While this employment reduction appears to be rather large, it was less than 5% of California's 1978 public employee workforce. And, of those 60,000 employees, layoffs accounted for 17,000 and the remainder resulted from attrition. Although there was a sharp increase in state financing of education to offset the reduced property tax revenues, some school program cuts, nevertheless, occurred. Particularly hard hit were summer school and adult education classes. Also, local governments in 1979 often found it necessary to reduce budgets for libraries, parks, and other recreational activities.

But these cuts were not as sharp as had been anticipated by opponents of Proposition 13. The reasons for this were twofold: many local governments instituted or increased user fees for certain public facilities to minimize closings or reductions in hours of operation, and state assistance to local governments was sharply increased. Because the increased assistance was made possible by California's large state surplus at the time Proposition 13 was passed, the California experience may be unique in its minimal effect on public service levels. Upon passage of Proposition 13, the state legislature provided $5.1 billion in "bail-out" assistance to offset local government revenue reductions through June 1979. When loans and property tax relief programs were subtracted from this amount, $3.7 billion was available for local government assistance. The funds were earmarked for education, selected health and welfare programs, and block assistance to cities and counties. A

142 URBAN GOVERNMENT FINANCE

second bail-out, passed in July 1979, provided $4.9 billion in assistance to local governments through June 1980. The state also will be financing 90% of nonfederal costs for Aid to Families with Dependent Children (AFDC), Supplemental Security Income (SSI), and Medicaid. This is particularly significant in light of the increasing costs of these programs in recent years. Between 1966 and 1977, state and local public welfare expenditures from own-source revenue increased from $3.2 billion to $17 billion, and expenditures for health and hospitals increased from $5.6 billion to $20.8 billion (ACIR, 1979).

Most states do not have the financial resources to offset substantial reductions in tax revenue to local governments. And it is unclear how long California can continue the magnitude of assistance provided in 1978 and 1979. The experience with Proposition 13 type of revenue restrictions is limited, and, indeed, it is difficult (and probably would be a mistake) to generalize from California's experience to other jurisdictions. While local revenue limits often require a rollback of taxes to prohibit revenue growth, state revenue limits are not generally as severe; revenue growth is permitted, though restricted. Sufficient time has not yet elapsed to appraise quantitatively the effects of spending limitations on state and local governments. But it is possible to make some judgments based on preliminary data and intuition.

The popularity of the limitation movement almost ensures that its effect will not be restricted to states passing controls, although these states will undoubtedly bear the brunt. Inevitably, revenue and expenditure limits on state and local governments will alter the relation among the local, state, and federal governments and will change the fiscal mix in the state and local budgets as well.

REVENUE LIMITS

As local property tax revenue growth is slowed or halted, local governments will have to significantly increase productivity, draw down available surpluses, or seek alternate revenue sources to minimize the disruption of public services.

To reduce costs, local governments may have public services performed by private contractors. The assumption underlying this is that costs can be reduced because the private sector can provide services more efficiently. Or the government may abrogate responsibility entirely to the private sector. In this case, the consumer would select the contractor and pay for the services directly.

The coming years may witness a proliferation of special districts to perform functions currently performed by local governments in states exempting special districts from limits. While special districts may circumvent the immediate problem, the proliferation of special districts can weaken local governments and make it more difficult to govern effectively.

Many California jurisdictions have discovered an alternate revenue source by imposing new and increased user fees for parks, pools, libraries, and other recreational facilities. Even new and increased fees, though they may prevent the closing of certain facilities, are unlikely to provide the revenue to offset the losses incurred by limitation. And imposing user charges or abrogation of public responsibility for providing certain services will have a differential effect on families and individuals with varying incomes. Lower income residents will bear a disproportionate share of the burden imposed by requiring direct payment for services and facilities.

Imposing new (or increasing old) rates of existing sales and income taxes are other possible revenue sources. While sales and income taxes are alternative revenue sources, only a bold public official or one who does not intend to run for office again would institute or increase these taxes on the heels of a resounding vote to reduce property taxes. In a quest for additional revenues, local governments will turn to higher levels of government for assistance. The need may be most compelling to offset revenue losses for education, as educational revenue is generally raised from property taxes.

In recent years, two phenomena have cross-cut federal and state assistance to local governments. State assistance to local governments has increased, thereby increasing centralization within the state and local sector at the state level. But local governments have received substantial aid directly from the federal government, thus fostering decentralization. This latter trend has provided local officials with increased discretion to raise revenue and determine expenditure allocations. In fact, the newest federal economic development program, the Urban

Development Action Grant, (UDAG) geared to reversing local economic problems, requires local government initiative, not necessarily accompanied by State assistance or concurrence, for grant approval. Local government officials have thus attained increased independence from the states as a result of Federal programs that provide funds and exclusive decision-making authority to substate officials. This independence may be sacrificed if local governments must now rely on state discretion to determine types and levels of local expenditures previously determined by local officials.

Decentralization is important because it ensures diversity and choice in public services and better response to the needs of a heterogeneous society (M. A. Levin, 1979). But there are those who would view an increase in the centralization of state-local relations at the state level as a positive outcome of the limitation movement. These individuals would argue that local governments are, in fact, progeny of the states and the primary responsibility for the fiscal condition of localities rests with the states and not the federal government.

Recently, state revenues were bolstered by the increases derived from sales and income taxes due to inflation. Despite these revenue increases, most states do not have the financial wherewithal to offset significant local government revenue losses and, thus, to prevent disruption of services. Increasingly, state governments are themselves subjected to either revenue or expenditure limits and will be unable to provide the large increases in intergovernmental assistance necessary to meet the local revenue shortfall over a sustained period. This is likely to become more severe as the decade progresses. Forecasters are projecting that most states will face slower revenue growth in the next 3-5 years than in the past 3-5 years, even without limitations (Bahl, 1979). Only in isolated instances, then, where states have accrued large surpluses, will they be able to provide California type of "bail-out" assistance to local governments.

Revenue limits on state governments combined with slower rates of state economic growth will not only affect states but will likely affect local governments as well. Unlike most local revenue limits, state limits are not net revenue reductions, but allow for specified levels of growth, generally tied to an economic indicator. If the limit does not permit real growth then, as in the case of local limits, state revenue limits will cause the states in which they exist to determine program trade-offs: which programs are to continue growing, which will remain unchanged, and which are to be reduced. If mandated costs, matching funds, and funding for programs that demonstrate a maintenance of effort to continue receiving certain federal funds are kept constant or increased, then other programs may have to be reduced, particularly if the effect of inflation is increasing costs more rapidly then revenues. To the extent that state and local governments have surplus funds, it is possible that rather than curtail services, the surpluses will be drawn down.

In the past, increases in federal intergovernmental assistance may have been counted on to rescue fiscally troubled state and local governments. But, the rapid growth in federal intergovernmental assistance that occurred in the 1960s and most of the 1970s has been slowed. During the 20 years from 1958-1978, intergovernmental grants grew at an average annual rate of 14.5%. Between 1979 and 1980, federal assistance to state and local governments is projected to grow at rates lagging the rate of inflation (Table 5.7). And recent efforts to fight inflation by balancing the 1981 federal budget are sure to result in real reductions in a number of federal aid programs and a further slowing of the growth in the intergovernmental grant system. This trend is likely to continue and, therefore, to increase the pressure on state and local governments to increase taxes, or, barring that possibility, to reduce services.

With pressure on Congress to balance the budget, the federal government will not intervene to prevent the disruption of services or employment due to state or local limitations. State and local governments are vulnerable to problems resulting from reduced growth in federal aid because of their increased dependence on this assistance. In 1950, federal intergovernmental assistance accounted for 14.2% of state and local government own-source tax revenue. By 1978, for each dollar of revenue raised from taxes, state and local governments received 40 cents from the federal government (Table 5.7). In their quest for additional revenues, states will not find the federal government receptive to their plight. While increases in federal intergovernmental aid are not expected to continue their previous rapid rate or even keep pace with inflation, as discussed

TABLE 5.7 Composition of Federal Grants-In-Aid (Current $s)

Year	Total Grants-In-Aid (millions)	Percentage Change Grants-In-Aid	Grants as Percentage of State and Local Own-Source Tax Revenue
1950	2,253	-	14.2%
1955	3,207	7.3%	13.7
1960	7,020	17.0	19.4
1965	10,904	9.2	21.3
1970	24,014	17.1	27.7
1971	28,109	17.1	29.6
1972	34,372	22.3	31.4
1973	41,832	21.7	34.5
1974	43,356	3.6	33.2
1975	49,834	14.9	35.2
1976	59,093	18.6	37.7
1977	68,414	15.8	38.9
1978	77,889	13.8	40.2
1979	82,858	6.4	NA
1980*	88,945	7.3	NA
1981*	96,312	8.3	NA

SOURCE: Special Analysis of U.S. Budget, 1981.
*Estimated.

above, the recent trend has been for the federal government to provide aid directly to substate governments. It seems unlikely, therefore, that States will receive increased federal largess in the coming years. Likewise, notwithstanding the recent increases in direct federal aid to localities, the outlook for continued increases is not encouraging.

If state and local governments cannot compensate for tax revenue losses through increased intergovernmental aid, services will be reduced. Additionally, state and local governments may find, due to revenue or expenditure limitations, they are no longer capable of demonstrating the necessary maintenance of effort or matching requirements for certain federal grant programs (e.g., UDAG or water and sewer grants), and thus may be disqualified from certain categorical aid. It seems reasonable to assume, then, that state and local governments will be reducing and deferring certain expenditures in the coming years. This does not augur well for capital expenditures and the funding of public employee pension plans, which, in the short run, tend to be invisible cuts. Already the deferment of capital expenditures and underfunded pensions are causing concern about their potential delayed impact (U.S. Joint Economic Committee, 1979; Peterson, 1978).

While pension costs have been growing rapidly in recent years, many jurisdictions have not been contributing adequately to cover the value of the pension benefits to members of the workforce. It is possible that, with additional budget pressures, pension systems may be further sacrificed to mitigate service level reductions.

Similarly, in austere periods, capital expenditures are among the first to be reduced. Between 1976 and 1978, there was an unprecedented 15.6% reduction in real capital expenditures by states (U.S. Bureau of the Census, 1976-1978). Similarly, local government capital outlays have been dramatically curtailed since 1975 (U.S. Bureau of the Census, 1976-1978). As a result, in many jurisdictions, funds are not available to replace aging facilities. Intuitively, state and local governments are not likely to reverse this trend, particularly when their budgets are further restricted by a combination of rising inflation and the limitation movement.

EXPENDITURE LIMITS

The effect of expenditure controls will not differ significantly from those of revenue controls. In most instances, revenue controls can be expected to affect expenditure increases, but may not do so in every case since expenditure limits are not mandated explicitly in all states with revenue controls. Since state and local government expenditure limits tend to exempt federal and state aid, it is possible that, if intergovernmental aid were to continue to show real growth, expenditure growth would continue relatively unhampered. In these austere times, however, state and local governments cannot count on increases in intergovernmental aid to maintain growth in their economy.

Expenditure limits provide for either a specified maximum percentage increase or an increase tied to an economic indicator. Because the local expenditure limit imposed in New Jersey is a specified maximum rate of increase (5%) that does not necessarily keep pace with inflation, without external assistance, services or employment will have to be reduced or efficiency and productivity increased unless certain spending categories are exempted from the limits. Arizona localities having a 10% growth limit since 1921 have increasingly pressured the legislature and courts to exempt more and more spending categories (Koon, 1978). As a result, about half of the typical local budget is exempted from the 10% limit (Koon, 1978).

Critics of the flat percentage curb argue that its rigidity prohibits expenditures from responding to changes in local needs, demography, and economic conditions. Supporters, however, believe costly new programs and public employee pay increases will be minimized and property taxes will be reduced as a result of specifying a maximum rate of allowable growth.

Local expenditure limits have been imposed in only two states in 1978 and 1979—California and Michigan—while state expenditure limits were adopted in ten. The more recent restraints imposed on state and local expenditure growth tend to be less restrictive. Rather than imposing a specific growth limit, expenditure growth is tied to the growth in a specified economic indicator—generally, the rate of growth in state personal income or the state economy. While this seems to indicate that expenditure growth will be slowed (because nine of the ten States imposing expenditure limits in 1978 and 1979 are in growing regions), it may still exceed the national rates of income and expenditure growth. And if inflation or a proxy for inflation is used as an indicator, expenditure growth may not be dramatically affected.

In some instances, "the rate of growth in the state economy" has not been precisely defined. In Hawaii, Tennessee, and Texas, this projection is made annually by the state legislature. Thus, there is room for discretion, and consideration may be given to extenuating circumstances requiring rapid growth. In Arizona, a three-member economics estimate commission determines the final estimate of total personal income for the following year, while in Michigan, an exacting definition is provided. Though the latter procedure removes the determination of "rate of growth" from the political arena, it also minimizes flexibility and is more difficult to alter.

If allowable expenditure growth does not keep up with the rate of inflation, we can expect services and employment levels to be reduced or efficiency increased. Correspondingly, this may permit own-source tax revenue to be reduced. To a large extent, however, this depends on the growth in "relatively uncontrollable" or exempted expenditures and revenue growth from external sources. If, for example, relatively uncontrollable expenditures, such as debt or pensions, increase significantly and revenues from external sources are reduced, it may not be possible to reduce own-source revenues.

CONCLUSION

Limits on local governments and a reduction in the growth of intergovernmental aid is likely to result in reduced levels of local government services and employment unless increased efficiency or productivity can offset the losses imposed by the limitation. While the limits on state governments are less restrictive than their local government counterparts, slowed state growth and reduced federal assistance may, in conjunction with the limits, result in reduced services and employment as well as a reduction in assistance to substate units of government. In instances where a budget surplus exists, these are likely to be drawn down before services are reduced.

Most of the states with limitations are in growing regions, and most state limitations do not set maximum limits on growth but tie the growth to an economic indicator. It is not unlikely, therefore, that taxes and expenditures in these States will still exceed the national rate of income growth. But this, may well be a reduction over the rate of growth that would have occurred without such limits.

Because the economies in the declining regions are slow growing, tying the allowable rate of increase in revenues or expenditures to the rate of state economic growth would have a dramatic impact. If the limitation movement becomes widespread, aggregate state and local government taxing and spending may slow, but the declining regions would experience the greatest reduction in growth (Bahl, forthcoming).

Ultimately, the limitation movement can be expected to minimize the implementation of new and the expansion of existing programs. Tax rate increases will be slowed, but user charges may prevent service cuts and sustain the flow of tax revenues to fund relatively uncontrollable budget items. Even if the limitation movement per se has lost some of its momentum, fiscal austerity at all levels of governments will prevail in the early 1980s. The biggest losers stand to be the poor and lower income families who may not be able to afford new or increased user fees or private services contracts.

APPENDIX A
State and Local Revenue and Expenditure Limitations, 1978

Revenue Limitations
　Local:
　　California—property tax rollback
　　Idaho—property tax rollback
　　Nevada*—property tax rollback
　State:
　　Michigan—tax revenue increase tied to growth of state personal income

Expenditure Limitations
　Local:
　　Michigan—local spending tied to state personal income
　State:
　　Arizona—State spending limited to 7% of total state personal income
　　Hawaii—State spending limited to rate of growth in state economy
　　Texas—State spending limited to rate of growth in state economy

　　Michigan—State spending tied to state personal income
　　Tennessee**—State spending limited to rate of growth in state economy

* This measure must pass again in 1980 before it can be implemented.
** Tennessee's spending limitation was passed in March, prior to the passage of Proposition 13.

SOURCE: See Appendix B

APPENDIX B
State and Local Revenue and Expenditure Limitations, 1979

Revenue Limitations

State:
Louisiana—State revenue lid tied to State personal income
Washington—Increase in state tax revenue limited to growth in state personal income

Expenditure Limitations

Local:
California—Spending tied to inflation rate plus population growth

State:
California—spending tied to inflation rate plus population growth
Nevada—State spending tied to 1975-1977 biennium budget adjusted for inflation and population change
Oregon—State spending ceiling for 1979-1981 biennium budget tied to rate of growth in state personal income
South Carolina—State is prohibited from spending more than 95% of current year's revenues; remaining 5% goes into a reserve fund.
Utah—State spending increases limited to 85% increase in state personal income

SOURCES: Shannon (1979); Austermann and Pilcher (1979) John Shannon, *Outline of Remarks before the National Municipal League,* Detroit, Michigan, November 13, 1979; and Winnifred M. Austermann and Daniel E. Pilcher, *A Legislator's Guide to State Tax and Spending Limits,* National Conference of State Legislators, March, 1979.

NOTES

1. Cited in D. R. Francis (1978) "Odd effects of California's Proposition 13 crop up," Christian Science Monitor, 27 June 1978.

2. For a comprehensive discussion of the causes and effects of the tax and spending limitation movement, see National Tax Journal Supplement 32(June 1979).

3. This does not assume the national average is optimum. It is used only for comparison.

REFERENCES

BAHL, R. (forthcoming) State and Local Government Finances and the Changing National Economy. U.S. Congress, Joint Economic Committee, Special Study of Economic Change. Washington: Government Printing Office.

--- (1979) Testimony Before the Subcommittee on Fiscal and Intergovernmental Policy. U.S. Congress, Joint Economic Committee. Washington: Government Printing Office.

COURANT, P. N., E. M. GRAMLICH, and D. L. RUBENFELD (1979) "The tax limitation movement: conservative drift or the search for a free lunch?" Discussion Paper 141, Institute of Public Policy, University of Michigan.

CITRIN, C. (1979) "Do people want something for nothing: public opinion on taxes and government spending." National Tax J. Supplement 32: 113-129.

KOON, B. (1978) "Living with limits." Wall Street Journal 3 November.

LADD, H. F. (1979) "An economic evaluation of state limitations on local taxing and spending powers." National Tax J. 31: 1-18.

LEVIN, D. J. (1979) "Proposition 13: one year later." Survey of Current Business 59: 14-17.

LEVIN, M. A. (1979) "Department of unintended consequences." Taxing & Spending 2: 15.

PETERSON, G. E. (1978) "Capital spending and capital obsolescence: the outlook for cities," pp. 49-74 in R. Bahl (ed.) The Fiscal Outlook for Cities. Syracuse, NY: Syracuse University Press.

RICE, W. N. (1978) State Tax and Expenditure Limitation: A Survey for Washington State. Olympia, WA: Washington Research Council.

SHANNON, J. (1979) "Tax limitation issues." Presented to the National Municipal League, Detroit, 13 November.

SHAPIRO, P., D. PURYEAR, and J. ROSS (1979) "Tax and expenditure limitations in retrospect and in prospect." National Tax J. Supplement 32: 1.

SIMMS, M. C. (1978) "Fiscal retrenchment: toward what end?" National Tax J. Supplement 32: 325.

SUYDERHOUD, J. P. (1978) "An analysis of the different tax limitation options," in Analysis and Alternatives to the Maine Tax Limitation Committee's Proposed Constitutional Amendment. Augusta, ME: Office of Legislative Assistants.

U.S. Advisory Commission on Intergovernmental Relations (1979) Significant Features of Fiscal Federalism. Washington: Government Printing Office.

——— (1977) State Limitations on Local Taxes and Expenditures. Washington: Government Printing Office.

U.S. Congress, Joint Economic Committee (1979) Deteriorating Infrastructure in Urban and Rural Areas. Washington: Government Printing Office.

U.S. Department of Commerce, Bureau of the Census (1976-1978) Governmental Finances. Washington: Government Printing Office.

6

The Location of Firms: The Role of Taxes and Fiscal Incentives

MICHAEL WASYLENKO
Pennsylvania State University

☐ THERE IS CONSIDERABLE DISAGREEMENT among economists and policymakers about whether state and local taxes and fiscal inducements, such as state direct loan programs and loan guarantees and local industrial aid bonds, influence interregional and intraregional business location choices. Economists have concluded that taxes and fiscal inducements have very little, if any, effect on industry locational decisions.[1] Thus, state and local policies designed to attract business are generally wasted governmental resources, since businesses that ultimately locate in a jurisdiction would have made the same decision with or without the fiscal incentive. While this conclusion is probably overstated, the burden of proof has now shifted to those who believe that fiscal incentives and taxation provide locational incentives.

Ancedotal evidence can be summoned to support either side of the issue. The purpose of this chapter is to explore the recent statistical evidence based on two types of analysis. The first is evidence based on analysis of interview data from surveys of firms, and the second is evidence based on econometric analysis using data from both secondary and primary sources. The

AUTHOR'S NOTE: *The author thanks the editor of this volume, whose comments led not only to editorial but also to substantive improvement of this chapter.*

evidence is examined for interregional and intraregional business locational decisions.

The first part of this chapter contains a profit-maximizing framework within which firms are assumed to make their locational decisions. The potential importance of taxes and fiscal incentives for locational decisions is discussed within this framework. The next section discusses the empirical findings about locational decisions and fiscal incentives. Some conclusions are contained in the last section.

The conclusion here is that taxes have only modest direct effects on business locational decisions, and that taxes are more important incentives for intraregional than for interregional choices. The results also indicate that not only the level of taxation but the distribution of state and local expenditures and tax burdens among income groups may influence business location and employment growth.

FRAMEWORK OF ANALYSIS FOR BUSINESS LOCATIONAL DECISIONS

THEORETICAL MODELS[2]

Locational theorists have developed models for analyzing optimal firm locational decisions. These theories point to cost of production and market demand at various locations as criteria. Weber (1929) developed a least-cost theory of manufacturing decisions. He recognized that raw materials and labor inputs were not uniformly distributed throughout any country. But he assumed that raw materials and labor are available at fixed locations, and that the product is shipped to a given market where it sells for a fixed price (perfect competition). The firm's problem is then solved in two stages. First, optimal production locations are selected by minimizing the sum of transportation costs of materials for production and the transportation costs for distribution of the firm's products to the market. The first-stage, optimal location is then modified to account for variations in labor cost and agglomeration economies among locations. The firm then locates at the point of least cost.

Lösch (1954) relaxed the assumptions of a fixed market and a given price. Population and demand, or markets, are not uniformly distributed over space. He noted that businesses would choose locations where their profits are maximized. These locations are not necessarily the points where transportation costs are minimized or where wages are lowest or where sales are largest. Profit maximization depends on the cost variation of all factors and the variation of market demand among locations.

Hotelling (1929) and Greenhut (1956) recognized the locational interdependence among firms. In these models, f.o.b. pricing is assumed, or it is assumed that consumers pay the marginal costs of distributing the product. In their models, there may be many firms selling the same product, but at any given location, firms establish a locational monopoly. Customers will purchase the product from the firm where the total price—price of the commodity plus transportation costs (financial and time)—is less than that of any other firm. If there is competition in the product market, the consumer will generally obtain the minimum total price at the nearest firm. Thus, once a firm establishes a location, competitive firms locating near that established firm reduce the area over which the established firm has a monopoly.

Greenhut provides an interesting and careful theoretical model of firm location. He recognizes locational variations in both the cost of factors used for production as well as the firm's demand or revenue. Moreover, demand varies not only by the size of the total local market but also by the number of other firms competing for the same market. The main point here is that a firm's profits vary by location. Moreover, the particular locational variables most decisive for a firm's choice vary by industry, since sales market and production inputs are likely to vary by industry. For example, firms in industries using specialized technical workers may locate exclusively based on the supply of technical labor, while firms using less specialized labor may choose locations based on other criteria. Spatial clustering of firms may also occur if localization or agglomeration economies exist in certain industries. Thus, the conclusions about factors affecting firm location can be expected to vary by industry.

EMPIRICAL MODELS

The theoretical literature and empirical studies of firm location are heavily oriented toward manufacturing locational decisions. But in principle, the theoretical and empirical models can be applied to nonmanufacturing locational decisions, although the decisive variables vary by industry. The differences among industries in the decisive variables are probably related to differences in the spatial distribution of relevant nonlabor inputs, labor skills and costs, and markets for products, as well as to differences in the cost of transportation for inputs and products among industries. Given the theoretical literature, a firm's decision depends on a vector of market (M) and cost (C) characteristics that vary by location.

The vector of market characteristics will vary with each industry. If the industry manufactures or supplies intermediate products, then the market variables will include the number of firms purchasing these intermediate products, the size of each purchasing firm's demand, and the number of competitive supplier firms at each location. If the firm produces for consumer markets, then the market variables may include the number of competitors, the per-capita income level of the market area, and the population size or the number of consumers at each location.

Cost factors include the supply of different types of labor (unskilled, skilled, and managerial); the cost of capital; the price and availability of land; proximity to a transportation network; transport costs for raw material and product distribution; agglomeration economies that reduce costs; energy prices; and the availability and cost of immobile inputs, such as bodies of water, coal, or other nonstandardized inputs.

Differences in state and local taxes could affect industry location in two ways. First, the direct effect of taxes would reduce the after-tax profits of firms, if taxes on capital (corporate income and property) are not shifted forward to consumers or backward to labor or even capitalized into the price of land. Even when taxes are shifted forward, differential taxes among locations would affect a firm's profits by raising the firm's prices and reducing their market area, unless the demand for the product is perfectly inelastic or demand is not responsive to price.

Fiscal inducements, such as state loan guarantee programs, direct loan programs, tax concessions, development credit corporations, and local industrial revenue bond programs, could also affect locational decisions. These programs reduce the cost of borrowing to firms and raise profits or increase the availability of capital.

Firms would of course be attracted to areas with lower taxes, other things being equal. The importance of taxes in the locational decision will depend on the size of the tax differential, on whether the tax differential between locations is substantial compared to differentials in other costs or markets among locations, and on whether the tax differentials result in higher (lower) quality or more public services for which firms and consumers are willing to pay. A firm may willingly pay higher taxes for public services that the firm would otherwise provide for itself—such as better drainage or more fire and police protection—in the absence of these public services. High-quality elementary and secondary schools and high taxes may also indirectly attract firms, since skilled and managerial labor may migrate to areas in part because of the quality of the educational system. The availability of skilled and managerial labor may, in turn, attract firms. Thus, firms may not always avoid high tax jurisdictions, especially if the high taxes are accompanied by higher quality public services that attract labor.

A potentially devastating criticism of locational theory is that firms do not search for optimal locations, but owners locate their firms on the basis of inertia, personal tastes for climate, and other "noneconomic" factors. Interview data often indicate that personal factors or historical accident, such as the location of the firm near the birthplace of its founder, are determining factors, especially among smaller firms. These conclusions are not necessarily inconsistent with locational theory.

The theory assumes a frictionless world, or neglects to include moving costs, profit uncertainties, and costs of information on new locations. In abstracting from these costs, it does not imply that they are unimportant. If firms had perfect information and moving was costless, then all firms would be at their optimal location. Firms do not have perfect information about costs and markets at all locations, and more than qualitative information may be costly for firms to obtain.[3] Firms move to a new location if the present discounted value of expected increases in future profits at a new location is greater than the moving costs; they also choose the location where future profits are highest. If information costs about alternate locations are the same for all, these costs will be a higher proportion of a small firm's profit differential between locations. Thus, information and moving costs may be reasons that proprietors of small firms stay in familiar areas or neighborhoods.

Locational theory also assumes that profits are the only basis for making decisions. Indeed, profits may be only one argument in a proprietor's utility function. Climate, tastes for urban or rural living, and other quality-of-life variables may be a dimension of a proprietor's interregional location decision. At an intraregional level, it is also often alleged that firms locate so that their managers can enjoy a shorter commute. These nonprofit arguments of the utility function are more important to a proprietor or partnership firm than a large corporation.

Locational theory can be modified to account for information and moving costs as well as preferences of owners and managers for locations with warm (or cold) climates and perhaps shorter commuting distances. But to the extent that the firm's profit-maximizing location is altered by personal preferences, the firm will tradeoff profits for personal factors. The owner is willing to forego a finite profit amount to satisfy nonprofit aspects in the choice of locations. Nonprofit dimensions of locational choice are not precluded here, but they are not the most important variables. Information and moving costs and a large fixed investment in a plant, however, may explain inertia.

Nearly all of the research on business locational decisions recognizes that firms do not necessarily locate in areas with the lowest wages, lowest taxes, or lowest cost for other inputs; nor do firms always seek areas with the largest product markets. The important locational determinants also vary by industry.

Some industries consider cost while others consider market. Moreover, firms in some industries may consider input; for example, some firms seek locations with an adequate supply of skilled labor. Such regions may be farther from markets and have higher taxes, but offer favorable climate and amenties attracting skilled and managerial labor. Others may locate in areas with less environmental regulation, greater energy availability, and more urban infrastructure, such as roads and water supply.

The locational effect of taxes may also vary by industry. Taxes on capital may not affect the location of firms in labor-intensive industries, but taxes on capital may affect the locational decision of firms in capital-intensive industries. Payroll taxes are more likely to affect the decision of labor-intensive firms than capital-intensive firms. And tax differentials and fiscal inducements are more likely to be decisive for firm location the smaller the area over which the location decision is being made. Simply stated, when cost and market differentials are small, as they are likely to be within a smaller geographic area, tax differentials are more likely to be the deciding locational determinant.[4]

A region's change in firms and employment is the net result of births, deaths, on-site expansions and contractions, and relocations of firms. To gain insight into locational determinants, researchers have explained four aspects of regional change in firms and employment; namely, regional variations in total employment, in the growth rate of employment, in the number of firm births, and in the number of firm relocations.

Results for locational determinants based on the number of firm births and the growth in employment due to firm births are the most reliable, since new firms have few moving costs and may be subject to less inertia. But proprietors of new firms may be reluctant to move far from their familiar surroundings, given the uncertainties about the success of a new business and the lack of information about locations other than their present ones. Relocating firms may have significantly more moving costs but are less uncertain about the success of their business and may also have lower information costs as a proportion of their profits. Thus, no definite preference for explaining either firm births or relocations emerges.

Firm on-site expansions and contractions as well as branch locations of multiplant operations (which can also adjust the size of their operation at each site) are also interesting areas for analysis. Birch (1979) indicated that regional variations in establishment on-site expansions and births are responsible for differences in employment growth among regions. By contrast, employment decline due to firm on-site contractions and deaths shows little variation among regions, while relocations of firms occur in relatively few cases. The relocation of the U.S. textile industry from the Northeast to the South during the 1950s is a fluke rather than an example of a typical industrial pattern.

In summary, using locational variables to explain regional variations in changes in total employment may mask important relations between firm location and the independent variables, unless all aspects of firm location in all industries respond to the same locational variables. Based on the theoretical literature, spatial variations in the supply of factors of production (especially labor), in markets for products, in agglomeration economies, and, perhaps, in taxes determine firm location. In some industries, where firms use nonstandard inputs, the availability of these inputs will constrain firm locational choices.

Information and moving costs may explain the firms' considerable inertia. Decisions may also be based in part on amenity variables, such as climate and other nonprofit variables. But amenity variables may attract a labor supply to an area and the labor supply may, in turn, attract firms. Whether firms follow a labor supply, or the labor supply is attracted to firms and employment opportunities or firms and labor supply are simultaneously determined is not addressed here.[5] Instead, the results of survey and econometric research on the locational variables that influence firm location are discussed below.

FINDINGS IN THE EMPIRICAL LITERATURE

INTERREGIONAL BUSINESS LOCATIONAL DECISIONS

Survey results. Surveying firms is one way to analyze firm locational decisions. The two most influential studies are those of Mueller and Morgan (1962) for Michigan manufacturing firms and Greenhut and Colberg (1962) for Florida manufacturing firms. The surveys asked the firms what was *most* influential in their locational decisions.

The Mueller and Morgan study surveyed new firms, relocating firms, and expanding firms. For the regional decision, new firms ranked the following as important: (in order of frequency) labor costs, proximity to markets, availability of skilled labor, industrial climate (defined as attitudes of the state and community toward industry), the tax bill, and proximity to raw materials.

When new firms were asked about the choice of a site within the region, 50% cited historically accidental or personal reasons (such as it was the birthplace of the founder or the owner has business connections there) for the specific site choice. Twenty percent cited a good business opportunity, and the remaining responses were related to cost or market factors. When new firms are disaggregated by size, smaller firms are more likely to cite personal reasons and chance, while larger firms tend to decide on purely economic grounds. Local tax concessions and fiscal inducements were not mentioned as most influential by more than 7% of the firms in any size category, but larger firms cited fiscal inducements as the main reason for their choices.

As Mueller and Morgan (1962) also note, small firms cite personal reasons because of familiar surroundings. Another interpretation is that small firms are more uncertain about the success of their business, and information and transaction costs are higher for these firms in unfamiliar surroundings. Small firms may find generating internal funds for investment more difficult, and local ties with familiar financial institutions may be especially important.

Relocating firms mentioned labor costs (32% of all firms) most frequently as the main reason for relocation, followed by taxes (20%), proximity to markets (18%), and availability of plant sites (10%). Plants expanding on-site mentioned proximity to customers (49%), followed by labor costs (23%), availability of plant sites (10%), and taxes (8%).

Mueller and Morgan (1962) also found a variety of reasons for inertia. Established business connections, the costs and uncertainties of relocation, and familiar places and people all insure that only substantial changes in profits will induce firm relocations. When relocations do occur, it is often in response to a decline in sales or profits. Moreover, taxes and local attitude toward business play a minor role in choosing a specific site. Greenhut and Colberg (1962) asked Florida manufacturing firms to cite various influential factors in their decision. About 22% cited access to product markets and another 15% cited anticipated growth of markets. About 11% cited low freight cost on shipping final product, and 8% cited climate as it affects operations.

Firms were then asked to cite the three main reasons for their choice. The most frequently mentioned were access to markets and anticipated growth of markets, ease of attracting skilled labor, climate, and community facilities for education, medical care, and police and fire protection.

While these surveys offer some insight into what determines a firm's choice, survey information is lacking on at least two grounds. First, the magnitude of the effect of each variable on location is determined only in a general way. Second, firms may use surveys (in part) to lobby for lower taxes or fiscal inducements and thus may overstate the importance of certain variables. One way to partially circumvent these problems is to observe what firms actually do and then econometrically relate their actions to a list of important variables. Alternatively stated, observe what firms actually do, not what they say they do.

Econometric evidence. Most of the econometric evidence on firm movement and employment growth is based on cross-sectional analysis. In this literature, regional differences in total firms and employment and the growth rate of firms and employment are related to regional differences in market and cost variables hypothesized to affect firm location. This literature is also heavily oriented toward manufacturing firm location choices.

Fuchs' (1962) study is perhaps the best known in this area. Fuchs examined changes in manufacturing employment from 1929 to 1954 among the states. He found that, as early as the 1930s, industry has been shifting to the southern and western states, and there was a comparative loss in industry in the northeastern states since 1929. Lower wages, warmer climate, less unionization, and lower population density were associated with increased growth in manufacturing employment. Fuchs does not consider taxes in his analysis. He argued that taxes are a small portion of a firm's costs, and tax differentials will not affect profits enough to offset other locational advantages. To further buttress his argument about the unimportance of taxes, he noted that higher taxes may be associated with more public services, and, thus, tax variations cannot always be considered an added cost to the firm.

Most recently, Carlton (1977a, 1977b, 1977c, 1977d) examined the importance of taxes and fiscal incentives. He examined variations in the births of single-establishment firms among U.S. SMSAs during the periods 1967-1971 and 1972-1975 using Dun and Bradstreet data. The estimates were performed on data from three industries: fabricated plastics (SIC 3079), communication transmitting equipment (SIC 3662), and electronic components (SIC 3679).

Carlton used a comprehensive list of SMSA variables to explain the births of firms. The variables include wages, supply of skilled labor, corporate and personal income taxes, property taxes, energy costs, proximity to markets and raw materials, unemployment rates, number of recent work stoppages, construction costs, land costs, and a business climate index. The business climate index is composed of business tax exemptions and other fiscal incentives, as well as state right-to-work laws, state minimum wage laws, state fair employment practice codes, and the presence of statewide industrial noise abatement codes.

For the plastics industry, more firm births occurred in regions where there were lower wages, more agglomeration economies, greater supplies of engineers (highly skilled labor), and lower energy prices for natural gas. More firm births in the electronics industry were associated with lower wages, higher rates of unemployment, more agglomeration economies, greater supplies of engineers, and lower energy prices. For the communication transmitting equipment industry, more births were associated with lower wages, higher unemployment, more agglomeration economies, and greater supplies of engineers. Energy prices are not a statistically significant determinant of firm births for this industry.

Among the statistically significant variables, lower wages and energy prices quantitatively had the largest effects on firm births. Taxes were unimportant; although the property tax variable did have a negative sign for two of the three industries, it was not statistically significant.

Carlton concluded that taxes and fiscal incentives were not important determinants of locational decisions, at least for these three industries. But his fiscal incentive variable did not adequately measure fiscal incentives—and for two reasons. First, fiscal incentives were included in a business climate index consisting of variables other than fiscal inducements. More importantly, the index measured whether or not states have a given fiscal incentive program and did not measure the extent to which fiscal incentives are offered or used by states and industries. Thus, Carlton's evidence on fiscal incentives is suggestive but not conclusive.

A recent study by Romans and Subrahmanyan (1979) expanded the inquiry about tax differentials and their effect on location to include tax levels, the progressivity of the tax structure, and the extent that state and local governments redistribute income through cash transfer programs. Specifically, the authors related the percentage increase in three variables—per-capita personal income, personal income, and nonagricultural employment in each state—to a vector of independent variables. These variables included the marginal tax rate in the state as a measure of tax progressivity; the ratio of state and local transfer payments to individuals corrected for unemployment variations to total state revenue as a measure of redistribution; the level of tax effort on personal taxes; the level of the tax effort on business taxes; average nonagricultural wages; percentage change in the region's per-capita personal income (personal income and employment) exclusive of the particular state; and the state's ratio of nonagricultural to agricultural income.

The results using employment as a dependent variable are the most interesting. They found that the fraction of state revenues used to finance individual transfers of income and the tax progressivity variables were inversely related to employment growth, and that the coefficients of these variables were statistically significant. But growth in the surrounding region, high business tax effort, and higher wages were directly related to employment growth; and the coefficients of these variables were statistically significant. Personal tax effort was also directly related to employment growth, but the coefficient was not statistically different from zero.

The positive coefficient for the business tax effort variable is particularly bothersome. The authors explain that the more industrialized states may be growing more rapidly and levying higher taxes on business. They also argued that the business tax effort variable may also partly measure the presence of public services for which business is willing to pay. But given that the coefficients of the wage and personal tax effort variable are also positive, it seems likely that their model was misspecified, and their estimates of the coefficients were biased. Their results, while suggestive, should be tested further before redistribution and tax progressivity can be said to influence firm locational decisions.

EVALUATION OF TAXES AND FISCAL INCENTIVES ON LOCATION

In this review, the assumption about tax incidence is that in equilibrium firms pay the average burden of taxes on capital, and that tax differentials are shifted to consumers, labor, or land. If firms are immobile, then they are more likely to bear the full burden of business taxes unless firms can shift taxes to consumers or factors of production without affecting their product market size or labor supply. This ability to shift taxes so easily seems unlikely if there is competition for market size among firms. Thus, the movement of firms may reflect firms adjusting from disequilibrium, caused partly by state and local tax differentials, to equilibrium.

The recent evidence, reviewed above, on taxation and interregional firm location decisions corroborates Due's (1961) earlier conclusions. Tax variables are more likely to be important in selecting a specific site within a metropolitan area, or when a suitable location is near a state border, than in the selection of a metropolitan area. Factor and product market considerations are more important determinants of interregional location choices.

This conclusion has several implications for state tax policy. It does not imply that states should disregard issues of business location in formulating tax policy. If state and local taxes in any state are sufficiently high, they may offset the cost or market advantages in the state. Some literature also suggests that taxes do not affect interregional location decisions, because state and local taxes do not vary significantly among states.

Morgan and Brownlee (1974) have examined the effect of taxation on variations in the implicit rental price of capital among Great Lakes' states from 1949 to 1970. They concluded that in any year tax burdens never vary more than four percentage points and generally vary less than two percentage points for these states. The U.S. ACIR (1967) suggests that states actually follow one of two approaches in making tax policy: a direct matching method or a tradeoff method. Under the direct matching method, states attempt to keep the burdens imposed by each of their taxes in line with the taxes of their neighboring states. Under the tradeoff approach, states offset a high tax rate for a specific tax with a low tax rate for another tax, so that their tax burden is in line with their neighboring states.

Tax burdens *do* vary among states. Table 6.1 shows the results of comparing total state and local tax burdens. Total tax burdens are compared using three published measures (U.S. ACIR, 1977): total state and local revenue from own sources as a percentage of state personal income; total state and local taxes as a percentage of state personal income; and per-capita state and local taxes. The mean value and the standard deviation of each of these measures for 50 states appear in Table 6.1. Table 6.1 also reports the low (and high) tax states, i.e., states whose tax burdens are more than one standard deviation below (or above) the mean, as well as the range of the burden measures.

Although there is some variation in these burdens, most states cluster within one standard deviation of the mean value. With the exceptions of Missouri and New Hampshire, the low tax states are in the South. Massachusetts, New York, and Vermont are all high tax states; New York has the highest state and local tax burden. Other high tax states are Alaska, California, Hawaii, Minnesota, Nevada, New Mexico, and Wisconsin.

The tax burden figures indicate that some high tax states,

TABLE 6.1 Variations in Tax Burdens among States

	State and Local Revenue From own Sources as a Percentage of State Personal Income	State and Local Taxes as a Percentage of State Personal Income	Per-Capita State and Local Taxes
Mean Value	15.6	11.9	$618.80
Standard Deviation	2.2	1.6	130.37
Low Tax States	Arkansas, Conn., Missouri, N. Hampshire, N. Carolina	Alabama, Arkansas, Florida, Tennessee	Alabama, Arkansas, Mississippi, N. Carolina, Oklahoma, S. Carolina
High Tax States	Alaska, Hawaii, Minnesota, Nevada, New Mexico, New York, Vermont	California, Hawaii, Massachusetts, Minnesota, New York, Vermont, Wisconsin	Alaska, California, Hawaii, Massachusetts, Minnesota, Nevada, New York
Range	12.7 (Conn.) - 23.0 (Alaska)	9.1 (Alabama) - 16.7 (New York)	$405 (Arkansas) - $1,025 (New York)

SOURCE: U.S. ACIR (1977: 46, 49).
NOTE: Low (high) tax states are identified as those that are more than one deviation below (above) the mean value.

especially western states, are among the states with rapidly growing employment. But New York, Massachusetts, and Vermont are at a disadvantage among the northeastern states because of their high tax burdens. Bahl (1979) and Mieszkowski (1979) present evidence that, in declining employment, New York State is the real disaster area of the Northeast. Of the two other high tax northeastern states, Vermont showed slight growth in employment in the 1970s; and the economic base of Massachusetts has been maintained. The latter was possible (despite declining employment in manufacturing) because the state's service employment has been growing, especially in Boston.

The case for high taxes to explain a declining economic base is not conclusive, since other high tax states show stable or increasing employment. Nevertheless, part of New Yorks' decline is attributable to its high personal and business taxes, inducing both the outward migration of population and business.

One study (Hanson and Touhsaent, 1978) examined business tax variation among 28 urban sites for two manufacturing industries: machinery (except electrical) and instruments and related products. For machinery except electrical, the effective tax rate varied between 10.8% in Randolph, North Carolina and 34.0% in San Francisco. For instruments and related products, the figures were 9.5 in Randolph and 29.3 in San Francisco. But the significance of these effective tax rate variations for firm location were not examined.

It is also argued that since tax differentials are not a large share of a firm's total costs, taxes are unimportant in the locational decision. But tax differentials may be a larger share of a firm's profits, and, thus, comparing taxes and profits may be more valid than comparing taxes and costs. Federal deductability of state and local taxes is often alleged to mitigate tax differentials. This is, in part, a bogus issue, since relative (the ratio of) marginal tax rate differentials remain regardless of federal deductability. But state and local taxes as a proportion of after-tax profits are reduced when state and local taxes are deductable from federal tax bases. Thus, federal deductability of state and local taxes reduces the absolute amount of the tax differential between regions.

The literature on interregional locational decisions is not conclusive. Several researchers (Carlton, 1977d; U.S. ACIR,

1967) suggest that the birth of branch plants may be more responsive to tax incentives than are other aspects of the business locational decision, since multiplant operations can locate capital intensive parts of their operations in low tax jurisdictions. But there are no explicit tests of this hypothesis.

The locational effects of fiscal inducements are more dubious, although only Carlton tested for fiscal inducement effects on firm location. But examining the cost of these inducement programs may provide some insight into the potential economic efficiency (benefits greater than costs) of inducement programs. Morgan and Hackbart (1974) examined the cost of state industrial tax exemption programs. Their estimates of the state revenues forgone as a result of the tax exemption programs imply that about one-half of the new industry in a state must be attracted by the tax exemption for the program's benefits (as measured by new industry value added) to exceed the program's cost. It is unlikely that one-half of the new industry of a state results from existing tax exemption programs. Thus, general industrial tax exemption programs will be economically inefficient subsidies. Fiscal incentive programs, such as industrial revenue bonds, applying only to new industries may be more efficient, since the costs of these programs are much lower.

In part, these fiscal inducement programs may not be effective because every state has at least two different tax incentive programs (Mulkey and Dillman, 1976). Bridges (1968) argued that the presence of tax concessions and fiscal inducement programs might improve a state's business image and at least allow the state to be a potential location. By lowering the price of capital, inducements could affect the volume as well as the location of investment. No evidence could be found on the extent that fiscal inducements affect the volume of investment, but fiscal inducement programs probably add more costs than benefits at the margin.

INTRAREGIONAL BUSINESS LOCATIONAL DECISIONS[6]

The distinction between inter- and intraregional locational decisions is not always an easy one. Movement between SMSAs as an interregional move and movement within an SMSAs as an intraregional move generally suffice as criteria. But when SMSAs are in close proximity to each other, such as New York and Nassau-Suffolk, it is not clear that a movement between these SMSAs is an interregional move. In this case, the distinction between interregional and intraregional is better made on a criterion of different product market areas or factor markets. For most SMSAs, interregional and intraregional moves can be categorized using the SMSA criterion.

If taxes bear on firm locational decisions, they are more likely to influence the choice of a specific site within a region than the choice of a region. ACIR (1967) and, to a lesser extent, Due (1961), both allow that taxes might be more important for a firm's specific locational choice. And when McMillan (1965) re-examined the survey or interview studies on firm location, he found that when firms are asked about their specific site choice within a region, taxes become a more important location determinant.

Despite these suggestions about the significance of local taxes, only a handful of studies attempt to determine the effect of taxes on intrametropolitan business locational decisions. The difficulty in obtaining data on firm locations, movements, births, and expansions accounts for the paucity of studies; almost all of the existing studies examine the location decisions of manufacturing firms.

The importance of local taxes rests on whether and the extent to which they vary across local jurisdictions. If local tax variations account for only a fraction of profits, then decisions are more likely to be based on factors other than taxes. Locational studies should consider variations in the total tax burden, or at least in the major local taxes, and not only in the property tax. In some metropolitan areas, jurisdictions levy only property taxes. Some jurisdictions levy income and property taxes, but the income tax rate is the same in all jurisdictions in the region, while the property tax rate varies among jurisdictions.

Often the unimportance of taxes is attributed to the fact that taxes are only a small portion of total costs. While local tax differentials compose a small part of total costs, they may be a significant part of profits. Suburban property tax rates are below those of their respective central cities (Morgan and Hackbart, 1974: 118). For Boston, Hamer (1973) estimated that the effective property tax rate for manufacturing firms is about 5% in the city and 4% in suburban jurisdictions. Stigler (1963) estimated that manufacturing firms earn a 14% rate of return on invested capital. If capital bears the full burden of the property tax—as it would if it were not mobile between city and suburb—a 1% property tax differential in the Boston area implies profit levels are 6% higher in the suburban jurisdictions than in the city. Thus, taxes are an important part of firm profits, especially when other markets and cost factors are the same across jurisdictions. Similar reasoning might apply to nonmanufacturing locational decisions if product and factor markets are the same between two jurisdictions.

The relocational decisions of central city firms can be viewed as two decisions. Existing firms decide to leave the city; and given that they decide to leave, certain variables determine their ultimate destination within the metropolitan area. Most studies have considered only one of the two aspects. The first stage of the decision is analyzed using survey data, while the latter is analyzed using survey data and econometric techniques.

Survey results. Hamer (1973) cited a questionnaire study of Boston SMSA firms, performed in 1969 by the Boston Economic Development and Industry Commission. The results indicated that firms are quite mobile. Between 1960 and 1969, nearly one-half of all suburban firms moved to their new sites. The questionnaire asked firms considering moves and those not considering moves about the reasons for their decisions. The responses of city and suburban firms were analyzed separately. Forty percent of the city and suburban firms considering moving cited space availability and cost; one-third of city firms cited labor-related factors. The questionnaire did not ask specifically about local taxes.

One-fourth of Boston city firms not interested in moving most frequently cited high costs of moving their equipment and a desire to be close to existing clients. This is some evidence that moving, adjustment, or transaction costs cause some inertia. Suburban firms cited favorable space and labor related factors.

Schmenner (1978) used questionnaire data to analyze the movers' decisions for firms in Cincinnati and New England. He concluded that mover firms are smaller (fewer employees) than nonmover firms; that space-related reasons, particularly the desire to expand the plant, are most frequently mentioned for moving; and that mover plants are more independent of local customers and suppliers than nonmover plants.

TABLE 6.2 Summary of Regression Models Used to Study Tax Influence on Intrametropolitan Firm Location

Study	Moses and Williamson (1967)	Beaton and Joun (1968)	Schmenner (1975)	Fox (1978)	Erickson and Wasylenko (1980); Wasylenko (1980)
Dependent Variables	a. Manufacturing mover firm destinations per unit land area. b. No. of manufacturing firm expansions per unit land area.	a. Percentage increase in manufacturing employment 1958-65.	a. Proportion of SMSA manufacturing establishments (employment) in each taxing jurisdiction. b. Change in proportion of SMSA manufacturing establishments (employment) in each taxing jurisdiction. c. No. of establishments (employment) moving from central city to tax jurisdictions.	a. Amount of industrial land in a jurisdiction 1969.	a. Proportion of establishments that move from central city to each jurisdiction j between 1964 and 1974. Movement for seven industries is examined: construction, manufacturing, transportation, wholesale, retail, finance and services.
Independent Variables					
Fiscal	a. Dummy variable = 1 if suburban zone = 0 if city zone	a. Effective Property tax rate (T). b. Recent percentage increase in property tax rate ($\Delta T/T$)	a. Cental city effective prop. tax rate minus suburban jurisdiction rate (T). b. Income tax differential (Y) defined like property tax differential). c. Central city suburb differential in per pupil educational expenditures.	a. Effective property tax rate (TRATE). b. Per capita police and fire expenditure (BSERV).	a. Effective property tax (TRATE). b. Per capita police and fire expenditures (SAFE). c. Per capita sanitation and street services (SANST).
Land Market	a. Distance from CBD (L). b. Percentage commercial and manufacturing zoned land that is vacant (V).	a. Price of land (P). b. Percentage of industrially zoned land that is vacant (V).	a. Distance to center city (D).	a. Price of land (LPRICE).	a. Reciprocal of distance from CBD(1/DIST). b. Percentage of land in industrial use (PIND). c. Percentage of land in commercial use (PCOM). d. Percentage of land vacant (PVAC).

TABLE 6.2 (continued)

Study	Moses and Williamson (1967)	Beaton and Joun (1968)	Schmenner (1975)	Fox (1978)	Erickson and Wasylenko (1980); Wasylenko (1980)
Labor Market	a. Population Density in each zone (W).		a. Population density (N).		a. No. of residents employed in each industry (i) within media commuting distance (7 miles) of each suburban jurisdiction (Ei) [Different (Ei) for each industry].
Transportation	a. Percentage of land area used for transportation (T). b. Dummy variable = 1 if near highway, = 0 otherwise	a. Proximity to railroad (R). b. Proximity to freeway (F).	a. Dummy for railroad access (R). b. Dummy for expressway interchange (H).	a. Dummy variables for access to interstates (AHIGH) and, b. Rail transport (RAIL).	a. Dummy variables for access to interstate highway (DHIGH).
Agglomeration Economies	a. Percentage of land in manufacturing use (M).		a. Existing density of establishments (E) (only used in net change and mover analysis).		a. Ratio of employment in industry (i) in each jurisdiction (j) to total noncentral city employment in industry (i) (AGj).
Consumer Market					a. Per capita income (PCY) b. Population density (DEN).
Other		a. Capital Assets/unit of industrial land (K).		a. Capital to land ratio.	
Simultaneous Variables		Price of land.		TRATE, LPRICE, BSERV.	
Regions Studied	Chicago	Orange Co. California	Cincinnati, Cleveland, Kansas City Minneapolis-St. Paul.	Cleveland	Milwaukee

For tax rates, Schmenner (1978) found in Cincinnati and New England that only about one-quarter to one-third of the relocating plants actually move to new locations with lower property taxes. About 40-50% either move within the same tax jurisdiction or to locations in towns with similar tax rates. Another one-fourth move to jurisdictions with *higher* property tax rates.

Apiledo (1973) interviewed Arizona firms receiving local government industrial aid bonds in an attempt to assess the role of industrial aid bonds in a firm's decision to branch, expand, or locate in a jurisdiction. Industrial aid bonds are used in 29 states and are the most substantial form of local government financial inducement. Apilado found no evidence that the aid bonds provided incentives for the location or volume of firm investments. He also found that corporate "need" is not evaluated by Arizona's municipalities in awarding financial assistance to firms, and that Arizona's local governments rarely assessed the cost and benefits of this program. He found no reason to expand or continue this incentive program.

Econometric evidence: demand. Moses and Williamson's (1967) study of intrametropolitan manufacturing locational decisions in Chicago is the first that empirically related firm expansions and relocations to a vector of cost variables. The independent variables are summarized in Table 6.2. Using 582 transportation zones in the Chicago area, the authors analyzed a cross-section of firm densities (number of firms in a zone divided by land area in a zone) for moving and expanding firms. For relocations, they found that fewer relocations occur in zones farther from the central city, and the percentage of land in manufacturing use in the zone had a positive effect on the number of firm relocations. The latter variable is likely to be a proxy for agglomeration economies, since more manufacturing land use implies more manufacturing employment and firms. The other variables in the model are not statistically significant.

The authors found that fewer expansions occurred in zones farther from the central city. The percentage of manufacturing land use in the zone had a negative effect on firm expansions. The latter variable may proxy the space in the zone for on-site expansion.

They also studied expansions and relocations of firms for zones in different subareas of the SMSA. The percentage of land in manufacturing use or agglomeration economies, is directly related to the number of firm expansions and relocations in a zone. The results for other variables are not conclusive. The results for the distance from the central business district (CBD) variable are the most consistent across zones; distance is inversely related to firm movement and expansion, but its coefficient is statistically significant in only four of the eight regressions.

They used a dummy variable—equal to zero if the zone is in Chicago and equal to one otherwise—to proxy tax, zoning, and other fiscal differences between city and suburban zones. The coefficient of this tax variable is never statistically significant. In general, the coefficients of the independent variables used to explain firm movement and expansion were not statistically significant and the study was of limited econometric success; but the study did lead to further research.

Beaton and Joun's study (1968) explained the percentage change in manufacturing employment for the period 1958-1965 in 20 cities in Orange County, California. They simultaneously estimated two equations: one for employment growth in a city, and a second for the price of land in a city. The price of land was treated as an endogeneous variable, since they considered the price of land simultaneously determined with manufacturing employment growth (Table 6.2). Among other variables, they included two tax variables—the effective property tax rate and the recent percentage increase in the property tax rate—in each of their estimating equations. Their two-stage-least-squares estimate of the manufacturing growth equation indicated that the tax variables were inversely related to manufacturing employment growth, but that a higher price of land implied higher manufacturing employment growth. The result for the price of land is plausible if firms prefer the amenities offered at the higher-priced sites to the remoteness and lack of amenities at lower-priced sites.

The authors' price of land equation produced curious results for the property tax variables. Higher values of each of the tax variables implied a higher price of land instead of the expected negative effect of property taxes on land values. Moreover, when the manufacturing employment growth and price of land equations are solved simultaneously to obtain the total effect of higher taxes on manufacturing employment growth, higher values of the tax variables implied more, not less, growth in manufacturing employment. The positive coefficients for the tax variables in the price-of-land equation and the positive coefficient for the price of land in the manufacturing employment growth equation generated this unexpected result. Thus, the authors' estimates are unreliable; and since higher taxes should not lead to higher land values, their analysis should be questioned.

Schmenner's (1975) study examined three aspects of manufacturing firm location using establishment and employment data for jurisdictions within four metropolitan areas. Schmenner explained the existing pattern of establishment (employment) densities (number of establishment in a jurisdiction divided by land area of the jurisdiction), the net change in establishment (employment) densities, and the relocation densities of establishments (employment) moving from the central city. He analyzed two different periods, 1967-1969 and 1969-1971, and studied Cincinnati, Cleveland, Kansas City, and Minneapolis-St. Paul.

Schmenner examined separately four SMSAs and a pooled sample of the four SMSAs; he used both establishment and employment as dependent variables and used two time periods. Therefore, he estimated twenty (5 × 2 × 2) regressions for the analysis of existing density patterns, for the analysis of the net change in density, and for the analysis of relocation densities. As independent variables, Schmenner used three measures of fiscal differentials between the city and each of its respective suburban jurisdictions: property tax rate differential, personal income tax rate differential, and per-pupil education expenditures differential. Other independent variables included measures of rail and highway transportation facilities in a jurisdiction, distance to the central city, population density (a measure of labor supply), and existing establishment density in the jurisdiction as a measure of available agglomeration economies (Table 6.2). An independent variable is a significant determinant of a particular locational decision (pattern, net change, and relocation) if its coefficient is statistically significant in all 20 (or in a majority of) regressions for the locational decision being analyzed.

The results of the analysis (20 regressions) of the existing pattern of firm densities indicated that the coefficients of population density (coefficients were statistically significant in 16 of the 20 regressions), railroad access (8), highway interchange (4), and the property tax differential (3) were statistically significant in a portion of the 20 regressions. Population density, a proxy for the distribution of the metropolitan labor force, most consistently predicted the existing density pattern of firms. The transportation variables were next in importance, while the property tax differential was only an occasional determinant. Specifically, the coefficient of the property tax variable is statistically significant and has the hypothesized sign for Cleveland in the 1969-1971 employment equation and for Minneapolis-St. Paul in both the 1967-1969 and the 1969-1971 employment equations.

The net change analysis met with less success. Population density (3), highway interchange (3), existing firm density (2), and railroad access (1) were statistically significant with the hypothesized sign in a small portion of the 20 regressions.

For the mover equations, the coefficients of population density (7), existing firm density (3), highway interchange (2), property tax rate differential (2), railroad access (1), educational expenses differential (1), and income tax rate differential (1) were statistically significant and had the hypothesized sign in a portion of the 20 regressions. The coefficient of the property tax differential is statistically significant and had the hypothesized sign for Kansas City in the 1967-1969 establishment and employment equations. The coefficient of the income tax variable was statistically significant and had the hypothesized sign for the pooled establishment regressions during the 1967-1969 period.

Schmenner's results indicated that the income tax variable was not a significant determinant of firm location, while property taxes may be significant especially in Kansas City, Minneapolis-St. Paul, and, perhaps, Cleveland. But the results of the property tax were not robust. The statistical significance of the property tax coefficient depended, but not consistently, on the period being analyzed, whether establishment or employment measures of firm location were used and which aspect of location (i.e., pattern or net change or mover) was being examined. In part, the unimportance of property tax rates for firm location may result, because during the period of analysis these rates were similar between these cities and their suburbs. In all cases, the central city effective property tax rate was no greater than 2.0% while the suburban rates are not lower than 1.3%. Thus, for cases where there was greater property tax differential between city and suburbs, property taxes may be important.

Schmenner concluded that tax effects were second-level concerns. A low tax rate was no guarantee that industry will be attracted to the jurisdiction, nor was a high tax rate a certainty to scare off new industry. Labor supply, as measured by population density, was the most consistent determinant of firm location.

Supply side. All of the above studies analyzed only the firm's demand for space in zones or localities. As Oakland (1978) also noted, these analyses omitted the willingness of communities to supply industrial sites. Fischel (1975) and Fox (1978) suggested that some local governments rationally excluded firms from locating within their boundaries or supplied no industrial sites. They argued that residents of communities derived utility from environmental quality, public goods, and private goods. The presence of industry in a jurisdiction presumably resulted in larger increases in tax revenues than in expenditures, and permitted more consumption of local public and private goods at the expense of environmental quality. Communities with stronger preferences for environmental quality will require higher tax revenues from firms or larger increases in private and public goods to accept losses in environmental quality.

Some communities with strong environmental preferences may want to increase their property tax rate on firms to adequately compensate them for losses of environmental quality. But since the property tax rate must be applied uniformly to residential and nonresidential property, residents are constrained to trade increased public goods for losses in environmental quality and losses in private consumption. If communities find either or both of these tradeoffs unacceptable, then they are likely to zone very little or no land for industrial use. A similar environmental quality argument can be made for commercial property, although most commercial property is likely to have less detrimental environmental effects than industrial property. Nonetheless, some suburban communities do zone out large shopping malls because of their environmental effects. The argument here (and in Oakland's study) is that including jurisdictions that zone out firms in an analysis of firm site choice may lead to biased estimation results, especially with respect to fiscal variables. Empirical tests of the community supply of sites are not available, and Fischel provides only indirect evidence. One of the prerequisites to Fischel's theory is that fiscal benefits exist for communities that admit industry and commerce. To measure the fiscal benefits to residents from business location, he related two measures of different fiscal benefits to the presence of commercial and industrial property. First, he regressed the mean household property tax payment on a vector of variables including commercial and industrial property values, median family income, public school enrollment per household, and miscellaneous revenues per household. Second, he regressed the total school tax levy (a proxy for expenditure) per pupil on commercial and industrial property values, median income, enrollment, and miscellaneous revenues. His sample was 54 communities in Bergen County, New Jersey. He concluded that there are some fiscal benefits to communities from commercial and industrial property. By his calculations, about 70% of all commercial and 52% of all industrial property tax payments benefit residents either by lowering residential taxes or increasing educational expenditures (Fischel, 1975: 155).

Since some fiscal benefits exist, Fischel then used a regression analysis to determine whether a lower amount of commercial and industrial property in a jurisdiction is associated with higher median family income in a jurisdiction, using proximity to New York City and land area of the community as control variables. He reasoned that if environmental quality is a normal economic good then higher median income communities will choose more environmental quality and zone a lower percentage of land for industrial and commercial property than will lower income communities. Higher income communities are more likely to find that the fiscal benefits of industry do not compensate adequately for their lower environmental quality.

His results indicated that higher income communities have relatively more commercial establishments, while industrial property is found in lower income communities. These results implied that lower income communities get the industrial firms, while higher income communities get commercial firms that probably cause less loss in environmental quality than industrial firms. But why lower income communities get industrial firms is not clear from these results. It is still an open question whether higher income communities zone out, i.e., supply few sites to industrial firms, or whether industrial firms demand sites in low income communities. Environmental zoning policies, however, are plausible explanations for the resulting location of industrial sites in lower income areas, if environmental quality is a normal good.

Fox (forthcoming) and Wasylenko (1980) analyzed demand for sites and incorporated the supply side into their models. A fully specified model of site choice would consist of demand and supply relations for industry sites using the tax rate as the equilibrating variable, and other variables as shifters of the demand and supply curves. Since data on the supply of sites or zoning practices in each suburban jurisdiction were difficult to obtain, neither Fox nor Wasylenko were able to use a fully specified model for the supply of sites.

Fox regressed the amount of land in industrial use in each of Cleveland's suburban jurisdictions on land market, fiscal, and transportation access variables for these jurisdictions (Table 6.2). Using 43 suburban jurisdictions as observations, he found that the coefficients of fiscal and other variables are not statistically significant. To consider the supply side, he then omitted from his sample 19 jurisdictions having no land in industrial use. He reasoned that these jurisdictions completely zoned out industry. When the same regression was run using the remaining 24 suburban jurisdictions as observations, the coefficient of each of the fiscal variables had the expected sign and each was statistically significant. This evidence supports Oakland's contention that the supply side should be an important consideration in models of firm location, and that demand analyses not accounting for jurisdictions zoning out industry may produce bias and misleading results for fiscal variables.

Erickson and Wasylenko (1980) estimated a model of establishment location choice for movers from Milwaukee City to its surrounding 56 suburbs. They estimated their model for seven different industries: construction, manufacturing, transportation, wholesale trade, retail trade, finance, and services. Their independent variables represented land markets, labor markets, transportation facilities, agglomeration economies, fiscal characteristics and, for some industries, consumer markets in each jurisdiction. Retail, finance, and service establishments were presumed to raise profits by locating in areas with more consumer demand and lower costs. Establishments in the other industries were assumed to locate on the basis of cost.

They found that moving firms in every industry chose suburban jurisdications having a supply of labor within commuting distance of the jurisdiction and that were near other firms in the industry—presumably to capture agglomeration economies. Lower land prices (distance from CBD) were statistically significant for some industries, but the coefficients of the fiscal variables were never statistically significant (Table 6.2).

Following Fox, Wasylenko (1980) reestimated the model omitting jurisdictions appearing to zone out industrial or commercial firms. Using movers instead of total industrial land may give better estimates of the statistical parameters, since total industrial land use presents only an accurate picture of location incentives to the extent that industry is in its equilibrium location. Given the possibility of considerable inertia in the land market due to moving costs, examining movers would seem a better test of the importance of fiscal variables. Moreover, Fox's results may be biased, since he omitted labor market variations among jurisdictions.

Wasylenko estimated the model for six industrial categories; since transportation industry is zoned out of almost all jurisdictions, it was not analyzed. Using land use data, he reasoned that 32 of the 56 suburbs zone out manufacturing firms, 21 zone out construction and wholesale trade firms, and 24 jurisdictions zone out retail, finance, and service firms. The results for the subsample indicated that labor supply, agglomeration economies, and land market variables were again statistically significant determinants of location choice for all six industries. Moreover, the coefficient of the property tax variable had the expected sign and was statistically significant for manufacturing and wholesale trade firm choices. But the tax variables were not statistically significant for choices in the other four industries. Wasylenko reasoned that manufacturing and wholesale trade establishments were more sensitive to property tax rates than were other industries, because firms in the other industries may follow consumer markets and place less emphasis on fiscal characteristics, while manufacturing and wholesale trade firms are more concerned with cost.

Evaluation. The results of the two studies incorporating the supply of industrial sites into locational choice analysis indicated that fiscal variables were statistically significant determinants for location of manufacturing and wholesale trade, i.e., industrial, establishments. But Wasylenko's results indicated that tax variables, although statistically significant, were less important predictors of establishment location than labor supply and agglomeration economies. Thus, tax variables appear to secondarily influence locational choices.

The conclusion that tax variables have little influence on firm location is subject to at least one important caveat. Tax differentials, while not directly influencing industry location, may affect the movement of population and, thus, labor supply, which, in turn, affects industry movement. No empirical study explicitly estimated a more general equilibrium model of firm and population movement, although some empirical evidence on the relationship between population and industry movement does exist.

Using statistical causality tests, Steinnes (1977) examined whether industry follows people out of the central city or people follow industry or whether the relation between industry and population decentralization is simultaneous. Steinnes found that service and manufacturing jobs follow people, but that people follow the retail industry out of the central city. Cooke's study (1978) extended Steinnes' work. He used estimates of employment and population density gradient parameters as measures of decentralization and examined different SMSAs. Cooke also examined hypotheses about industries following each other and hypotheses about industries following people and concluded that jobs follow people. His results indicated that services follow retail, which in turn, follows manufacturing, which follows people: Industry follows a labor supply (or market) out of the central city. But why do people leave the central city? If it is due partly to fiscal factors, and industry *is* following people, then fiscal factors indirectly affect industry movement.

The classic urban economics literature assumes that the demand for land is income elastic, and, thus, people migrate to suburban areas to consume more land more cheaply in spite of the added transportation costs associated with suburban residence. Specifically, the utility gained from the additional land consumption in the suburbs compensates for the lost utility associated with additional commuting from the suburbs to the city center. To explain why families in higher income classes move to the suburbs, this literatuie assumes that the demand for land is more income elastic than is the income elasticity of the marginal utility lost as a result of commuting. Thus, families in higher income groups consume more land and commute farther than their lower-income counterparts.

Recent evidence rejects this classical theme. Bradford and Kelejian (1973) concluded that families at or above middle income migrate to the suburbs, in part, because of an unfavorable central city fiscal residual, i.e., expenditure benefits for these income groups are less than the taxes they pay. Wheaton (1977) examined the land transportation cost tradeoff explanation for population movement. His evidence indicated that the income elasticity of land was quite low compared to the disutility of extra commuting. Wheaton concluded that the demand for land does not explain the migration to the suburbs. Instead, people migrate from the central city to the suburbs due to favorable fiscal factors and other amenities such as educational quality in the suburbs.[7]

Given this evidence, taxes may indirectly affect the movement of industry from the central city. Further research using a more general analysis relating industry and population movement to fiscal and other variables is necessary before the role of fiscal variables in firm location can be ascertained.

Another worthwhile refinement for firm locational research may be more disaggregation of industries, since all industries are not subject to the same cost, input, and market variables. Kemper (1975) examined a sample of firm births in five areas (CBD, central industrial district, core, inner-suburban ring, and outer ring) of the New York City SMSA between 1967 and 1969. For each zone, he related the percentage of the births in each manufacturing industry that occur in that zone to a vector of industrial characteristics. He argued that input requirements for the industry and input prices are important for locational choice.[8] He found that industries using more nonstandard inputs, such as water, more professional labor, more unskilled labor and manufacture products for consumer use (as opposed to intermediate products), were more likely to choose sites in the central city sites. Thus, not all industries migrate to the suburbs nor are they all responding to the same incentives.

CONCLUSIONS

Taxes and fiscal incentives play little or no role in a firm's choice of locations among regions. This results partly because market and cost variables vary more among regions than do fiscal variables. The lack of variation in fiscal variables occurs because states follow a direct matching method or a tradeoff method in making tax policy. Specifically, under the direct matching method, states keep each of their tax rates in line with those in neighboring states. Under the tradeoff approach, states keep their overall tax level in line with that in neighboring states but do not match each specific tax.

Nonetheless, some states have exceptionally high state and local tax burdens. But only New York has suffered severe economic decline in the 1970s. Thus, while tax burdens vary among states more than the location literature attests, empirical evidence that taxes affect interregional business location decisions is almost nonexistent.

It does not follow, however, that a state should ignore business location in making tax policy or that a state should unilaterally eliminate fiscal inducements because they are ineffective. Although fiscal effects on firm location are limited, the lack of any fiscal inducements in a state may cause firms not using fiscal incentives to believe erroneously that the state has a bad business climate. Firms view a state as being fiscally unpredictable (bad business climate) in its policy toward business if the state appears unconcerned with tax and fiscal inducements for business.

The intraregional evidence on the effect of taxes is less definitive. When firm locational models take account of a community's supply of industrial sites, taxes are a statistically significant determinant of industrial locations. Still, taxes are secondarily important for locational decisions. Although taxes may not directly affect industry location, there is some evidence that taxes indirectly affect industrial movement, since taxes may influence the decentralization of population from central cities, which, in turn, influences the movement of industry. No research to date has estimated an intraurban model of population and employment decentralization. The insignificance of taxes for firm locational decisions may result from considering only the direct influence of taxes on business location. More careful analysis of the interaction of firm and population movement, especially in an intraurban context, may more sharply delineate both the direct and indirect influence of taxes on decisions.

The research may be refocused toward firm births and expansions. Birch (1979) and Birch et al. (1979) found that employment growth occurred in regions and in jurisdictions within a region because of differential numbers of firm births and on-site expansions in the region and jurisdiction. Differential numbers of firm deaths, on-site contractions, and relocations did not vary significantly among regions or among jurisdictions within a region. Differentials in employment growth cannot be attributed to variations in the latter three aspects of firm location. Research should examine patterns of firm births and on-site expansion.

For central city policy toward retaining industry, survey evidence from Schmenner (1975; 1978) and Hamer (1973) suggested that many firms leave because of their inability to acquire enough land to expand their operations. The authors' conclusions suggested that cities interested in retaining industry carefully use their power of eminent domain to clear land for firm on-site expansions. Nonetheless, the movement of industry from central cities does not respond to many policy variables and, while it may be possible to slow industrial decentralization, it is not possible to stop it. And evidence from Kemper (1974) suggested that some industries are more likely to locate in the central city than the suburbs, especially industries using professional or unskilled labor and manufacturing products for consumer markets. Thus, some industries are likely to remain in the city and to be attracted by central city sites in spite of the city's fiscal disadvantages.

The economic function of cities has changed. Central city locations will not favor most manufacturing industries; instead, cities will continue to attract industries using a high proportion of professional workers, such as business services and related commercial industries. Central city tax policy may alter the pace of this functional change, but taxes will not alter the economic forces dictating the functional change in the industrial structure of cities and regions.

NOTES

1. For a review of the earlier literature, see Due (1961).
2. For a more detailed summary, see Miller (1977).
3. The Simmons Gordon Publishing Company's "Plant Location" contains data on wages, labor availability, market data, climate, fiscal incentives and taxes, energy prices, and other information for individual states and their SMSAs. This information is not costly, but not specific enough for most firms to calculate their profit change as a result of a move.
4. An extreme but illustrative example of the influence of tax differentials on firm location within a small area is the town of Lloydminster, Canada. Lloydminster is located partly in Alberta, a conservatively governed province with lower taxes and lower minimum wages, and partly in Saskatchewan, a socialist province with high taxes. Most businesses locate in the Alberta section because of their low taxes and lower minimum wages. Alberta also has a personal income tax rate, while Saskatchewan has larger transfer payments to poor families. As a result of these fiscal policies, most of the poor in Lloydminster live in the Alberta, while the middle class live in Saskatchewan (Zehr, 1980).
5. But for an excellent review, see Vaughn (1977).
6. See Oakland (1978) for a review of some of this literature.
7. Aronson and Scwartz (1973) also provided some evidence that fiscal variables determine intraurban migration patterns. They analyzed the fiscal residuals and patterns of migration for jurisdictions surrounding Harrisburg, PA, and argued that the population in jurisdictions with favorable fiscal residuals should grow faster than those with less favorable fiscal residuals. But their analysis did not take into account other economic factors affecting migration.
8. Since taxes are uniform in the central city zones, he did not include taxes in the analysis.

REFERENCES

APILADO, V. P. (1973) "Public administration of financial incentives in industrial plant location: industrial aid bonds." Papers in Public Administration 26, Arizona State University, Institute of Public Administration.

ARONSON, J. R. and E. SCHWARTZ (1973) "Financing public goods and the distribution of population in a system of local governments." National Tax J. 26: 137-160.

BAHL, R. (1979) "The New York State economy: 1960-1978 and the outlook." Occasional Paper 37, Metropolitan Studies Program, Syracuse University.

BEATON, C. R. and Y. P. JOUN (1968) The Effect of the Property Tax on Manufacturing Location. Fullerton, CA: Division of Real Estate, California State College.

BIRCH, D. (1979) "The job generation process." U.S. Department of Commerce, Economic Development Administration. Washington, DC: Government Printing Office.
–––, E. S. BROWN, R. T. COLEMAN, D. W. DaLOMBA, W. L. PARSONS, L. C. SHARPE, and S. A. WEBER (1979) "The behavioral foundation of neighborhood change." U.S. Department of Housing and Urban Development. Washington, DC: Government Printing Office.
BRADFORD, D. and H. KELEJIAN (1973) "An econometric model of the flight to the suburbs." J. of Pol. Economy 81: 566-89.
BRIDGES, B. Jr. (1965) "State and local inducements for industry," parts I and II, National Tax J. 18: 175-192.
CARLTON, D. W. (1977a) "Locational decisions of manufacturing firms." Report 7728, Center for Mathematical Studies in Business and Economics, University of Chicago.
––– (1977b) "Births of single establishments firms and regional variations in economic costs." Report 7729, Center for Mathematical Studies in Business and Economics, University of Chicago.
––– (1977c) "Models of single establishment births." SIC 3079, Report 7730, Chicago: Center for Mathematical Studies in Business and Economics, University of Chicago.
––– (1977d) "Models of new business location." Report 7756, Center for Mathematical Studies in Business and Economics, University of Chicago.
COOKE, T. (1978) "Causality reconsidered: a note." J. of Urban Economics, 5: 538-542.
DUE, J. F. (1961) "Studies of state-local tax influences on location of industry." National Tax J., 14: 163-173.
ERICKSON, R. and M. WASYLENKO (1980) "Firm relocation and site selection in suburban municipalities." J. of Urban Economics 8: 69-85.
FISCHEL, W. (1975) "Fiscal and environmental considerations in the location of firms in suburban communities," in E. S. Mills and W. E. Oates (eds.) Fiscal Zoning and Land-Use Controls. Lexington, MA: D. C. Heath.
FOX, W. (forthcoming) "Fiscal differentials and industrial location: some empirical evidence." Urban Studies.
––– (1978) "Local taxes and industrial location." Public Finance Q. 6: 93-114.
FUCHS, V. (1962) Changes in the Location of Manufacturing in the U.S. Since 1929. New Haven: Yale University Press.
GREENHUT, M. L. (1956) Plant Location in Theory and Practice: The Economics of Space, Chapel Hill: University of North Carolina Press.
––– and M. R. COLBERG (1962) Factors in the Location of Florida Industry. Tallahassee: Florida State University Press.
HAMER, A. M. (1973) Industrial Exodus from the Central City. Lexington, MA: D. C. Heath.
HANSON, E. and S. TOUHSAENT (1978) "State and local taxation in 1977: a comparative analysis." Rochester, NY: Center for Governmental Research.
HOTELLING, H. (1929) "Stability in competition." Economic J. 39: 41-57.
KEMPER, P. (1975) "Manufacturing, location and production requirements." (mimeo).
LÖSCH, A. (1954) The Economics of Location, New Haven: Yale University Press.
McMILLAN, T. E., Jr. (1965) "Why manufacturers choose plant locations vs. determinants of plant locations." Land Economics 41: 239-246.

MIESZKOWSKI, P. (1979) "Recent trends in urban and regional development," in P. Mieszkowski and M. Straszheim (eds.) Current Issues in Urban Economics, Baltimore: Johns Hopkins University Press.

MILLER, F. W. (1977) Manufacturing: A Study of Industrial Location, University Park: Pennsylvania State University Press.

MORGAN, W. D. and W. E. BROWNLEE (1974) "The impact of state and local taxation on industrial location: a measure for the Great Lakes region," Q. Rev. of Economics and Business (Spring): 67-77.

——— and M. M. HACKBART (1974) "An analysis of state and local industrial tax exemption programs." Southern Econ. J. 41: 200-205.

MOSES, L. and H. WILLIAMSON, Jr. (1967) "The location of economic activity in cities." Amer. Econ. Rev. 57: 211-222.

MUELLER, E. and J. N. MORGAN (1962) "Locational decisions of manufacturers." Amer. Econ. Rev. 52: 204-217.

MULKEY, D. and D. L. DILLMAN (1976) "Location effects of state and local development subsidies." Growth and Change (April): 37-43.

NETZER, D. (1966) Economics of the Property Tax. Washington, DC: Brookings.

OAKLAND, W. H. (1978) "Local taxes and intraurban industrial location: a survey," in G. Break (ed.) Metropolitan Financing and Growth Management Policies. Madison: University of Wisconsin Press.

ROMANS, T. and G. SABRAHMANYAN (1979) "State and local taxes, transfers and regional economic growth." Southern Econ. J. 46: 435-444.

SCHMENNER, R. W. (1978) "The manufacturing location decision: evidence from Cincinnati and New England." Report to U.S. Department of Commerce, Economic Development Administration.

——— (1975) "City taxes and industry location," revision of his unpublished Ph.D. dissertation, Yale University (1973).

STEINNES, D. (1977) "Causality and intraurban location." J. of Urban Economics 4: 69-79.

STIGLER, G. (1963) Capital and Rates of Return in Manufacturing Industries. New York: National Bureau of Economic Research.

U.S. Advisory Commission on Intergovernmental Relations (1977) "Significant Features of Fiscal Federalism 1976-77 Edition, Vol. II Revenue and Debt," Washington, DC: U.S. Government Printing Office.

——— (1967) State-Local Taxation and Industrial Location. Washington, DC: Government Printing Office.

VAUGHAN, R. (1977) "The Urban Impacts of Federal Policies: Vol. II, Economic Development," Santa Monica: Rand Corporation.

WASYLENKO, M. (1980) "Evidence on fiscal differentials and intrametropolitan firm location." Land Economics 56: 339-349.

WEBER, A. (1929) Theory of the Location of Industries, Chicago: University of Chicago Press.

WHEATON, W. C. (1977) "Income and urban residence: an analysis of consumer demand for location." Amer. Econ. Rev. 67: 620-631.

ZEHR, L. (1980) "Life can be so sweet on the sunny side of a boarder street: but which Lloydminster side, Alberta or Saskatchewan, is really the sunny side?" Wall Street Journal 5 March.

7

The Next Decade in State and Local Government Finance: A Period of Adjustment

ROY BAHL
Syracuse University

☐ THE 1980s WILL BE A PERIOD of fiscal adjustment for state and local governments. The formerly rich states will be struggling to bring their *relative* quality of public services down to a level they can afford; the formerly poor states will be struggling to raise service levels in response to the demands of their new populations; and all will be trying to adjust to a higher rate of inflation and a slower growing U.S. economy. The lessons on getting along with less will painfully be learned by more than a few state and local governments.

How will changes in the U.S. economy affect state and local government finances in the 1980s, and what governmental policy responses will be necessary? To answer these important questions, we first consider those national economic and demographic factors that may shape the outlook, then discuss the essentials of a national urban policy and of the possible adjustments by state and local governments. We conclude with a guess at what the next few years in state and local government finance will hold.

AUTHOR'S NOTE: *This chapter is an expansion of Chapter 6 of my "State and Local Government Finances and the Changing National Economy," prepared for the Special Study on Economic Change of the Joint Economic Committee.*

FACTORS SHAPING THE OUTLOOK

That state and local governments everywhere are facing problems of adjustment is a reflection of the changing structure of the U.S. economy. A slowing national income growth and a shift in its regional distribution, a continuing high rate of price inflation, a changing population structure, changes in federal budget and federal grant policy, and a new voter resistance to big government and regulation, all exert important pressure on the financial condition of state and local governments and all call for some form of policy response. In truth, the changes are less recent than some policy analysts should be willing to admit—the slower rate of income and population growth has been recognized for several years, as has the ongoing pattern of regional shifts in population and economic activity. But old fiscal habits die slowly, and adjustments take time. The growth in government is just beginning to slow and the realities of long-term retrenchment are only now taking hold in some jurisdictions in the declining regions. The reverse is true in the growing regions where increasing costs and the pressures to upgrade services are beginning to affect state and local government budgets.

NATIONAL ECONOMIC GROWTH

The prognosis for the 1980s is for real GNP to grow more slowly than in the 1960s and 1970s. Between 1970 and the first quarter of 1980, real GNP growth was positive in seven years and averaged 4.5%. For the ten years of positive growth rates in the 1960s, the average was 4.1%. Certainly the next two years will not begin to approach this rate. The administration has projected a real GNP decline in 1980 and a real growth of only 2.0% in 1981 (Congressional Budget Office, 1980).

Few will hazard outright projections of GNP five or ten years in the future, but some indirect evidence casts doubt on the believability of 4% to 5% real growth rates for the early 1980s. The administration estimates that to achieve a 4% unemployment rate by 1985 and a 3% inflation rate by 1988, annual productivity increases of 2.5% and real GNP growth rates in the range of 4.5% to 5.0% will be required. To the extent these long-term inflation and unemployment targets are not attainable, slower real income growth will result.

The Bureau of Labor Statistics made baseline projections of a 3.2- to 3.6% annual real growth rate in GNP for the 1980s. These projections require that inflation slow to 5.5% in the early 1980s and to 4.4% by the end of the decade and that the unemployment rate gradually fall from a projected level of 5.3% in 1981 to 4.5% by 1990 (Saunders, 1979: 12-24). The Congressional Budget Office (1980) has simply assumed (calculated) a 3.8% growth rate "so that by 1985 the unemployment rate would return to approximately the current level (5.9 percent)" [1980: 2-5]. The Joint Economic Committee (1980: 30-32) assuming productivity increases in the 1.5- to 2% range, sees the long-term rate of real GNP growth to be in the 3- to 3.5% range. From almost every vantage the conclusion seems to be the same. For at least a few years, the U.S. economy will grow more slowly than it did during the past two decades.

One important reason why the more optimistic scenarios such as the real-growth targets set by the administration may not be reached is that the inflation rate will likely remain high in the 1980s. The underlying causes of inflation have been building for more than a decade and cannot be swiftly corrected—indeed, the President's 1980 *Economic Report* recognizes this in pushing back its timetable for lowering the rate of inflation. Moreover, some major causes of inflation are a result of world events—oil pricing, production decisions, and crop failures—and are neither controllable by domestic policy nor predictable. The prospects for easing price increases in the 1980s might also be viewed in terms of the components of inflation. The major contributors in recent years have been energy, housing, food, and medical costs. Neither federal policy nor international events would cause us to expect a dampening in any of these components of general price increase. Reischauer's review (this volume) of forecasts supports this pessimism—he expects the 1980s to be characterized by relatively slow economic growth, high rates of inflation, high levels of unemployment, rapid nominal wage growth, and high interest rates.

This combination of slower real growth and inflation will put new pressures on the budgets of state and local governments.

Forecasts for the state and local government sector are not generally available, though the Bureau of Labor Statistics' (BLS) projection model is an exception. Under their baseline employment expansion assumptions, they expect the sector to decline between 1980 and 1985 in employment (12% of total employment to 11.6%), purchases of goods and services (12.6% of GNP to 11.1%), and personal taxes (3.2 to 2.9% of GNP) [Reischauer, this volume]. Whether or not the relative declines in state and local government activity will be this steep, it would seem reasonable to assume that taxes will be off their post-1975 annual real growth rate of 4.3%. If the past few years is representative and if tax limitation movements do not further slow tax revenue growth, a 3.5 to 4 percentage real GNP growth could imply a state and local government tax revenue growth of 2.7 to 3.1% per year.

The resulting revenue gap will not likely be made up by increased federal assistance. To the contrary, if the federal grant share of GNP remains constant, a 3.5- to 4.0 percentage real GNP growth will bring an annual increase in Federal grants of 4.6- to 5.3%. Even this projection, which seems optimistic, is for a growth well below the 7.3 percentage annual real increase of the period 1975-1978.

The import of all this is clear. State and local governments will have less resources available in the 1980s—the overall rate of revenue increase could fall by as much as one-fourth if the real GNP growth rate stays in the range of 3.5- to 4%.

Some areas will be hit harder than others by this slow national growth and by the cutbacks in the real amount of federal aid to state and local governments. The slower-growing industrialized states in the Northeast and Midwest could experience very little real growth under this scenario and central cities in those regions will be the hardest pressed. Governments in this region could well face revenue growth rates lower than the national rate of inflation—a combination of slow, real national growth and declining regional shares. Many of the growing states will not escape from the revenue effects of the national slowdown. Those growing states without substantial energy resources will face a more drastic reduction in their *rate* of revenue increase than will many of the Northern states who have already entered a period of fiscal austerity.[1]

The other side of the coin is inflation, and to some extent inflated tax bases offset the dampening effects of slow economic growth. But property taxes are not so responsive to inflation, and continued inflation and taxpayer resistance will eventually force rate reduction or indexation for more state government tax systems. These factors will probably hold back inflation-induced revenue growth so that it will not offset the losses due to slower growth. The more significant effects of inflation on state and local government budgets are likely to occur on the expenditure side. If the pattern of recent years holds, rapid increases in costs will account for most if not all of state and local government expenditure increases. This implies little or no increase in the real level of services offered.

Higher rates of inflation also promise two structural changes in state and local government spending. The first is that with soaring material and supply costs, a more labor-intensive public sector might seem feasible. The clamor of the past decade for increased productivity by capital-labor substitution may diminish in favor of arguments for more policemen and fewer cars and the like. The other major structural change is the extent to which capital formation in the state and local government sector will slow even further. Rising material costs, rising interest rates, and the ease of deferring the renovation and maintenance of the capital stock could all contribute to further reducing the rate of investment by governments in renewing their infrastructure.

REGIONAL SHIFTS IN ECONOMIC ACTIVITY

The slowing of national economic growth will be more than offset in some regions by the inmigration of economic activity. In the older, declining regions, it will be reinforced. There are prospects for people and jobs moving to the newer region, a trend that should continue through the end of the century. Estimates of regional populations and income growth by the Department of Commerce (Water Resources Council, 1974; Bureau of Economic Analysis [READ], 1977) and regional population and employment growth by the Oak Ridge Laboratory (1977; Olsen et al., 1977) agree. Census population projections offer a similar prognosis (Bureau of the Census, 1977).

But no matter how sophisticated the model, the projections are an extrapolation of past trends and will not pick up major turning points. One might question whether there are factors at work beginning to slow these regional shifts.

Evidence of a new equilibrium? There is some evidence and logic to argue that the growing and declining regions are approaching a new economic equilibrium. One line of argument would consider the limits to growth in some parts of the Sunbelt—water and the ability to provide services to accommodate a large population increase. Another would consider the relative cost of doing business. Labor costs may now be growing as fast in the South as in the North, and there is some evidence that the overall cost-of-living is rising faster in the South. Weinstein (1979) reports that between 1972 and 1978, the BLS's level of living index rose by 66.4% in southern cities in the sample but only 56.6% for cities in the Northeast. A continuation of this differential rate of price increase will drive up relative labor costs in the South and could be reinforced by increasing union strength—a natural consequence of manufacturing moving to the newer regions. The increasing cost of Sunbelt living may improve the attractiveness of northern plant locations, but the convergence is painfully slow.

One might speculate that the rate of taxation is becoming similar and therefore will slow regional job shifts.[2] This would be little more than speculation. Tax burdens have not become more uniform across the 50 states, though a few high income-high taxing states have cut taxes or slowed their rate of growth relative to personal income, while some low-taxing states have increased effective tax rates to fill backlogs of unmet services and respond to increasing population and income. For example, the declining states of New York and Ohio reduced their relative tax burdens from 1975 to 1977, while growing states such as California and Colorado had relative tax burden increases. Yet, for the most part, the declining states had relative increases in tax burdens and the growing states had relative declines. This result is not at all inconsistent with a slowing rate of increase in taxes in high-income states—the problem is that financial capability grew even slower. The reverse was true for many of the growing states—they did not increase taxes fast enough to keep pace with growth in their taxable capability.

The effects of the energy crisis on regional shifts in economic activity are anything but clear, but the net effect may well accelerate the decline. The prospects for higher energy prices and uncertain supplies in northern and midwestern states suggest a bias in the locational decisions of energy intensive firms toward the growing regions. And rising energy prices can produce a bonanza in energy tax revenues for some state governments. This could substantially ease any fiscal pressures on those states and remove one bottleneck to their continued growth. On the other hand, the rising cost and more limited use of air conditioning could deter southern economic growth.

Two other factors argue against regional convergence. One is that markets have shifted away from the older regions, and to the extent jobs follow people, the job share in the declining regions may still have a way to go. Finally, there is the question of consumer taste or relative preferences for northern versus southern living. The current pattern of migration would suggest a comparative advantage to states that can offer more sunshine and less congestion.

There may indeed be forces operating to slow regional shifts by raising the comparative advantage of the older industrial states. If so, these turning points are too recent to be detected. A more likely prospect is for a continuation of the Sunbelt shift of the 1970s.

Fiscal adjustments. Regional movements of population and economic activity will pressure state and local governments to adjust their fiscal behavior. For some northern states the scenario will be continued, long-term retrenchment. As a state like New York attempts to bring per-capita expenditures (40% above the U.S. average) into line with per-capita income (4% above the U.S. average) the central issue becomes how to *lower* the level of public services relative to other states. Few states, and especially New York State, have experience with such matters.

Such an adjustment is not only slow, but it is complicated by a number of factors:

- Inflation is driving up costs faster than revenues, accentuating real service level declines.

- Slower real income growth cuts into an already thin margin of revenue coverage.
- Many northern states are characterized by highly decentralized fiscal systems, hence it is difficult for the state government to plan for or control the aggregate level of state and local government spending and taxing.
- Because of jurisdictional fragmentation, the fiscal position of central cities in the declining regions is likely to be hurt a great deal more than that of suburbs, i.e., much of the costs of retrenchment are ultimately paid by low income families.
- There are important psychological barriers to retrenchment—residents find it much easier to adapt to lower taxes than to adapt to lower public service levels.
- The strength of public employee unions, fixed debt and pension commitments, a backlog of needed infrastructure improvements, and the existing near-crisis financial conditions of many cities make substantial retrenchment especially difficult.

The net result of all this is that while regional shifts in economic activity demand that the formerly rich states bring their fiscal activities into line with their new, relatively low levels of income, the retrenchment probably involves a period of public sector atrophy in the North. This means that governments probably will not and cannot cut back service levels in the absolute, but if they do not raise tax burdens or expand the quality and quantity of services and spend just enough to keep real per-capita expenditures approximately constant, in time the rest of the country will catch up. This is long and slow and implies making public service levels *relatively* worse, but it is the kind of adjustment most likely to occur.

The growing regions will also face fiscal adjustment problems. On the one hand, there is great rural poverty in the South and Southwest and the need to use substantial amounts of the revenues from growth to deal with these problems. Then there are the pressures from growing population and income to expand infrastructure, improve school and health systems, deal with water shortages and environmental problems, and control land use. The growing regions would seem more equipped (than most northern States) to deal with these pressures for a number of reasons:

- Resources are growing because of regional shifts, even though national growth is slowing, and because state tax structures in the growing regions tend to be more inflation-sensitive than those in the Northeast and Midwest.
- Governmental finances tend to be more state dominated and therefore more controllable.
- Many urban areas are not characterized by fragmented local government structures.
- Some states will experience substantial revenue growth with rising energy prices.

On the other hand, there are state and local government financial problems ahead for southern states. Much of this increase in spending could come in the form of a catch-up in average wages, hence expenditures may rise more rapidly than public service levels. Employment levels relative to population are already higher in southern than northern states, as are levels of per-capita debt.

DEMOGRAPHIC CHANGES

Major changes in the national demographic makeup will continue through the year 2000. Fertility rate reductions and mortality rate declines have combined to push the nation toward a zero population growth, an increasing concentration of the elderly, and a declining proportion of school-aged children. Concomitant with these trends has been an increasing rate of household formation. The potential effects of these changes on state and local government finances could be significant. Unfortunately, this is a virtually untouched research area, hence we can but pull together some disjointed evidence and speculate about fiscal implications.

Expenditure effects. A slower population growth has uncertain implications for productivity, labor force participation, and the growth in GNP, hence the implications for state and local government revenues are uncertain.[3] But a slower population growth rate would seem to imply less pressure on the expansion of public services and therefore less pressure on public budgets. For some services, this is intuitively clear. Education, roads and streets, and water and sewer services quickly come to mind. Yet the situation is considerably more complicated. First, the questions must be carefully framed. How does a slower versus a faster rate of population growth, *ceteris paribus*, affect state and local government finances? What are the fiscal implications of slower population growth for particular jurisdictions and for the aggregate financial position of the state and local government sector?

A lower (rather than a higher) national growth rate might be translated into actual population declines in some older regions and central cities. On the surface this would alleviate some severe budgetary pressures. Yet the literature is uncertain about the effects of changing population size on public expenditure levels. Consider first the growing cities and states. Despite a great deal of discussion about the possibility of scale economies in the provision of local public services, there is little or no hard evidence to suggest that larger cities could deliver services any more cheaply on a per-person basis than could smaller cities (Bahl et al. 1980). One would conclude from this that a greater rate of population growth, *ceteris paribus*, means a greater increase in expenditures. Conversely, the loss of city or state population does not guarantee an expenditure reduction because there are many offsetting factors, e.g., inflation, mandates, and the simple creation of excess capacity in the city plant. Muller (1976: 82-83) has shown that per-capita common function expenditures between 1969 and 1973 for 14 declining cities rose by 51%, as opposed to 59% for 13 growing cities. As a percentage of personal income, he found the growth to be even greater for the declining cities. The determinants of public expenditure change are far too complicated to allow any precise estimates of the cost savings of a slower population growth rate. We can guess that an increase in the rate of population growth, *ceteris paribus*, increases expenditures and vice versa. But we do not have a feel for the magnitude of that effect in different types of jurisdictions.

If the question is whether slower population growth, *ceteris paribus*, reduces the aggregate level of state and local government spending, the answer is probably that it does. A faster population growth would not only generate more service demands but it could stimulate more migration.[4] The *movement* of population, as much as the size of population, increases costs, i.e., servicing a new suburban population may increase public sector costs by a greater amount than the cost reductions resulting from outward migration from an old neighborhood. While differential rates of population growth may have significant budget effects, the more important effects on public expenditures are likely to come from the changing composition of population. The compositional changes most important in this respect are the increasing proportion of the elderly, the declining number of school-aged children, declining urban densities, and declining urbanization.

A growing elderly and retired population could affect public budgets by causing shifts in social service expenditures and by putting pressure on the financing of retirement needs. The two most likely areas of concern are retirement cost and health care expenditures, though other public assistance programs may also be affected. The pressures brought by an older population on social service expenditures by state and local governments may not be so severe as one might expect. State and local governments spend substantially more on health care for the elderly than for the younger age groups, but less than 9% of total state and local government expenditures are for health and hospitals and about 85% of health expenditures on the elderly are aided. Moreover, one interesting set of projections suggests that growth in the numbers of elderly will be offset by growth in their income (from earnings and social security) leaving the proportion eligible for public assistance essentially unchanged over the next 40 years (Goodman, 1979). A potentially more important pressure on state and local government budgets may come from the problems of financing state and local government pension plans. If a government were operating on a pay-as-you-go basis, or with substantial unfunded liabilities, and if the age distribution of public employees changed in the same fashion as the demographic makeup of the community, then taxes to finance retirement cost expenditures could rise substantially in the 1980s (Munnell, 1980).

There is a bit more evidence, albeit indirect, on the expenditure effects of other types of compositional changes. Empirical work suggests that declining population densities may reduce spending for urban services such as police and fire, and a falling pupil-to-population ratio could eventually lead to lower educational expenditures (Barro, 1978). As welcome as such relief might be, one should not think too quickly about the possible uses of such savings. First, the effects of inflation may more than offset any "quantity" reduction, and anyway there will be substantial adjustment costs associated with budgetary shifts, i.e., such as from youth to age-related programs. Other "compositional" factors might offset the savings from a slower rate of population growth. The formation of new households will bid up certain costs—sanitation and fire—and the continuing movement of population to suburban and nonmetropolitan areas may cause the unit costs of providing public services to rise.

Revenue effects. The changing growth rate and composition of population will also be felt on the revenue side of state and local government budgets. The subject has not been thoroughly worked and one cannot go to a developed body of literature to support speculation about how changing demographics will change revenue flows. Still, an increasing *share* of the elderly will dampen revenue growth if for no other reason than because of an income effect. Retirees earn less and therefore have less to spend on taxable state and local government items—taxable consumer goods and housing. A related hypothesis is that a dollar of retirement income does not generate the same amount of tax revenue as a dollar of wage and salary or proprietorship income. The elderly receive special relief from state taxes through property tax circuit breakers, their housing choices run toward less expensive housing, and they consume a greater share of income in nontaxable housing, food, and medical care.

Another compositional factor is that the ratio of dependent to productive age group individuals will decline through the mid-1980s but then begin to increase with increases in the elderly and those under-10-years of age. Hence, the rate of growth in real sales and income tax revenues could be dampened by the late 1980s.

The other demographic change with important fiscal implications for state and local governments is the changing number of households. A taste for smaller families, the divorce rate, the postponement of marriage and childbearing, and the declining fertility rate have slowed the rate of population growth but not the formation of households. An example of the magnitude of this effect is New York State where a 9% increase in population is projected between 1980 and 2000, but a 25% increase in households (N.Y. State Economic Development Board, 1978). The fiscal implications have not been carefully studied. At first blush more households within a given population size implies more income earning units and therefore more taxable capability. More property *units* would suggest a bouyancy for the property tax, taxable income should increase, and there should be an increase in the taxable consumption share of income. The counterargument is that more young families may result in an increased stock of lower valued housing units, and there may be relatively little effect on the property tax. The expectation that more household units will increase the aggregate marginal propensity to consume taxable items (because younger families will go into debt to increase their purchase of durables) is debatable at best.[5]

Overall budgetary implications. A priori, the fiscal effects of a changing rate of growth and composition of population are so unclear as to be inconsequential, except perhaps for the costs of adjusting budgets to the new mix of services required. Yet, because some regions will realize these demographic changes more than others, more substantial fiscal effects could emerge. The increasing proportion of the aged and the increasing number of households is a national phenomenon, but the slower rate of national population growth is not being felt to the same extent across all regions. A continuing interregional migration will compensate for declining birth rates in some regions and reinforce natural population decline in others. Particularly the central cities will feel the change in becoming older and smaller but with an increasing number of households. If the fiscal consequences of demographic change turn out to be harmful, it is these cities that will be hurt most.

THE LIMITATION MOVEMENT

It is not likely that the tax revolt movements of 1978 and 1979 signal a permanent reversal in the growing share of government in GNP. But it seems clear that fiscal limitations of one kind or another will be a significant influence on state and local government budgets during the next five years. By mid-1979, 30 state legislatures and the U.S. Congress were considering balanced budget amendments. Some 14 states passed some form of tax or expenditure limitation between 1978 and 1980 (see Matz, this volume). The mood is clearly in the direction of slowing the growth of government at all levels.

The explanations of this dissatisfaction are many (Burkhead, 1979). Increasing taxes would be especially objectionable during inflationary times, when real spendable earnings for most American families have hardly increased. As long as the rate of inflation remains high, the objections from this group of voters will remain substantial, and growth in government will be resisted. In particular, rising property tax rates place onerous burdens on homeowners in that accrued worth may differ markedly from annual income. Shapiro et al. (1979) argued that the high and rising property tax burden was at the heart of the Proposition 13 movement. Yet Matz (this volume) pointed out that limits have been adopted in states *not* experiencing high or rapidly escalating taxes or expenditures. Another source of discontent is what is perceived of as an inefficient public sector—overpaid, underworked, and not responsive to citizen needs. Whatever the reasons for this dissatisfaction, it seems likely that some state and local governments will be tied to personal income growth in what they are allowed to spend.

Fiscal limitations, if they stick, will reduce the discretion of government decision-makers in formulating new programs and taxes and in altering the timing of their own fiscal expansions and contractions. Even though there is an option to switch to user charge financing (a compensating device used in the aftermath of California's Proposition 13), it is clear that local fiscal planning will be more constrained, and new spending initiatives will be bypassed to meet increased spending for "less controllable" budget items.

It is less clear what the effects on aggregate state and local government fiscal activity will be. On the surface, tying tax and expenditure growth to personal income growth suggests a dampening effect. Yet 13 of the 14 states imposing such limits are in the growing region—only Michigan is a declining state. Hence, even with limitations, a growth in taxes above the national rate of income growth could occur (though one might speculate that it would be even higher without the limitation). Moreover, in nearly every case the limitations apply only to state government. In total, the affected governments account for no more than one-fourth of total state-local government revenue raised from own sources. It is difficult to see how the limitations per se would significantly hold down aggregate state and local government spending. And, even with state tax limitations it is not clear that local spending and taxing would be slowed. The ACIR argues that it would, by 6- to 8% per capita by comparison with nonlimitation states, while Ladd disagrees (Advisory Commission on Intergovernmental Relations, 1977: Ladd, 1979).

On the other hand, if there were a more widespread adoption of such limitations, aggregate state and local government taxing and spending would slow but by a significantly greater amount in the declining region. In some states this discipline would be welcome, but it does reduce fiscal flexibility in states where fiscal capacity is growing more slowly.

Perhaps a more significant effect on the budgets of state and local governments is the possibility of limitations at the federal level. The proposals range from a fixed maximum percentage increase in federal outlays to a ceiling on the ratio of federal outlays to GNP. But all would slow the growth in Federal spending (see Reischauer, this volume). Even without a legal indexing of federal expenditures, the tax revolt movement will press to balance the federal budget more frequently than ever. Some of this balancing will inevitably result in reduced resources available for the more controllable federal grant-in-aid programs and in a further dampening effect on state and local government revenues.

The limitation movement at all levels of government gained some momentum in 1978 and 1979, and still more states will probably adopt varying controls on their budget growth. But inflation, public employee wage demands, federal assistance cuts, and slow economic growth will eventually catch up with some limitation states and stall the limitation movement in others. The limitation and austerity concerns of this year could give way to a renewed worry over deficient public service levels by the mid-1980s.

State legislatures will eventually reason that limitations will not address the underlying problem of an inefficient public sector that so rankles many taxpayers, nor is it clear that it will stimulate local economic development as others hope. Further, limitations may cause state and local governments to make revenue-raising adjustments such as increased use of benefit charges and the creation of special districts. Such policies may well be in the public interest under many circumstances, but not likely if their adoption is justified as a way around a formal limitation. The adjustments by state and local governments to circumvent debt limitations, and the efficiency and controllability of the resulting agency arrangements, is a lesson worth remembering.

Limitations are not without virtues. They force the political process to accept the fate of allowing a government to live within its means. Yet this discipline is accomplished at a cost of substantial flexibility in fiscal decision making and may induce some inefficient behavior by the limited government.

REVITALIZATION

Some analysts and many journalists see a revitalization of central cities taking place. It is not usually made clear whether revitalization means increased city population, employment and income, an improved economic position of the central city relative to suburbs, or simply a physical rehabilitation of certain parts of the inner city. Some, who borrow the term "gentrification" from the British, see it as filtering housing (or neighborhoods and retail districts) upward from working class to professional middle class.[6] Whatever the meaning, the implication is that the inner cities of the future will be much less the distressed areas that they now are and that federal policy toward cities ought to be adjusted accordingly. Indeed, some public policy is premised on the ability to induce more employment and residential activity in depressed inner city areas. A national development bank and tax abatements for construction investments in blighted areas are good examples.

The revitalization argument is based on *a priori* reasoning, casual observation, and wishful thinking. It has several elements.

First, changing demographics may favor central cities over suburbs. More singles, childless couples, and elderly in the national population; the increased demand for rental housing, smaller and less-expensive housing; and the convenience of city living (mass transit, convenience for shopping, and so forth) will bring people back to the city. And the deterrent of poor public schools in central cities will be less important for families without children. Second, the energy crisis will favor the city. Workers will move closer to work—and perhaps to where mass transit is available—to avoid the longer and more expensive commute. Third, there is the "bright lights of the city" argument. With more cultural and social activities, cities are exciting places to live, and some new awareness of these benefits will bring back white collar, middle income workers. Finally, there are the agglomeration effects which make the city a competitive location for certain types of white collar and service businesses. As evidence of revitalization, proponents give many examples: A booming Manhattan, Chicago's loop, and Capitol Hill.

Accepting the revitalization arguments as a basis for policy-making is better than wishing on a star. But not much. There is little evidence that city populations are increasing, that—relative to suburbs—their income and employment levels are rising or that their disadvantaged are better off. Indeed, none of these patterns have materialized. Central cities declined in population by about 5% between 1970 and 1978, they declined as a share of metropolitan area population and employment, and the city/suburb per-capita income disparity has actually grown (see Fossett and Nathan, this volume). If there has been a back-to-the-city movement, it has been dwarfed by the effects of those factors stimulating decline. Even the a priori arguments on revitalization seem flawed. There is some appeal to the notion that childless couples and singles see the city as a desirable location, because they are not deterred by poor quality schools and because of proximity to amenities and work. Yet the *postponement* of having children does not necessarily mean that couples will remain childless or that children will not be planned for. Indeed, some have argued that the fertility rate in the United States will soon increase. If this occurs, the quality of the public schools remains a major drawback to city residential location choices. Locations closer to amenities may also be a comparative disadvantage of cities, e.g., most cities cannot compete with the convenience and choice of suburban shopping centers, and the mass transit system is a major inducement in only a few cities.

The energy argument may also be questioned. There are more suburban than central city job locations, hence if the rising price of gasoline induced any population movement, it may well be to suburban locations. Moreover, if the commute to work grew too expensive, other kinds of adjustments might be made: e.g., a four day work week or innovations in communications to minimize necessary personal contact. To the extent movement took place in response to commuting costs, it would likely be blue-collar manufacturing workers moving to suburbs. Some white-collar workers might be lured to the city, but again the quality of the public schools would be an important impediment.

The bright-lights argument is based on a notion of cities being exciting centers of cultural and social activity that make city living more exciting. The impression is true enough, perhaps, for a Manhattan or a Georgetown but would hardly seem to fit a Syracuse or Toledo.

This is not to argue that revitalization is undesirable, that cities should not be brought back. Rather it is an argument for care in defining revitalization and for realism in assessing what can happen in cities during the next decade. Revitalization can mean a conservation of capital facilities, reinvestment in blighted areas, and a general improvement in the quality of city life. This pattern would be perfectly consistent with shrinking population and employment, the displacement of the poor from dilapidated housing in rundown neighborhoods, and the continued loss of manufacturing employment. Revitalization of cities, in this sense, may be a reasonable expectation. But it will not mean a diminished need for federal help in compensating for the economic losses, subsidizing the disenfranchised, and generally getting through a tough adjustment period.

FEDERAL POLICY

The federal government will play a major role in getting state and local governments through the difficult period of fiscal adjustment ahead. The question is whether the federal response will be reasoned and comprehensive or ad hoc and piecemeal. Some general guidelines for the federal response must be worked out, i.e., the kind of strategy one might expect to find in a well-thought-out statement of national urban policy. In its absence, some rough generalizations about how such a policy *might* view the financial problems of state and local governments is offered here. They fall into four areas of question about the appropriate federal response to urban problems; whether the federal government ought to attempt revitalization of declining areas or compensation during a period of financial adjustment; whether inflation and recession ought to be viewed as a part of intergovernmental policy; what role should state governments play in the intergovernmental system, and what will be the Federal policy toward possible big city financial disasters.

COMPENSATION VS. REVITALIZATION

If the administration's urban policy statement of 1978 took any firm position, it was toward a revitalization rather than a compensation strategy (Office of the White House Press Secretary, 1978). The National Development Bank, the targeted employment tax credit, Neighborhood Commercial Reinvestment programs and expanded UDAG funding all seemed to lean toward renovating a deteriorated economic base in distressed cities. At least the rhetoric of federal policy would imply a belief that the declining economies can be revitalized. Yet there is little evidence that such programs work or have any effect on the employment base of declining cities.

A policy of compensation would take a different tack by accepting the notion that market forces are affecting a reallocation of population and income within the country. It would attempt to compensate the most financially pressed governments and families caught in this transition. The goal would be to protect particularly the low income by subsidizing public services and temporary job opportunities while the "emptying out" goes on. Public service job programs, categorical grants in the health and education area, and federal relief for welfare financing would be key elements.

There is a fine line between revitalization and compensation, and we should not confuse the latter with any program to abandon cities or declining regions. As interregional variations in the relative costs of doing business and in market size approach some new balance, movements in population and jobs will slow. A primary role of federal policy is to assist the most distressed governments during the adjustment. Hence, subsidies to hold businesses in a region are not an appropriate part of a compensation strategy if the business will leave (or cease operations at present levels) when the subsidy is removed. "Transition" grants to states (such as New York) with an overdeveloped public sector are appropriate if they are tied to longer-term reductions in the level of public sector activity. Capital grants to renew the city's infrastructure are also appropriate, if the infrastructure investment is based on a "shrinkage" plan. Finally, relocation grants and labor market information systems are perfectly consistent with such a strategy, because they facilitate the outward movement.

THE BUSINESS CYCLE AND INTERGOVERNMENTAL POLICY

The business cycle and inflation have dramatic effects on the financial health of state and local governments. Indeed, the severity of the last recession pushed New York City over the edge and brought many other local governments and at least one state dangerously close to fiscal insolvency. Because swings in economic activity induce substantial changes in fiscal health, an explicit recognition of business cycle effects in federal intergovernmental policy is necessary.

In a sense, this was done with countercyclical aid and the stepping up of other components of the economic stimulus package in the last recovery, but it was done ad hoc rather than as part of a coordinated federal, intergovernmental policy. The basic objectives of the Comprehensive Employment and Training

Act (CETA) were training and employment of the disadvantaged and then countercyclical stimulus. Local public works were meant to stimulate state and local government construction. Some would argue that both became general purpose fiscal relief programs, and that neither stimulated the economy (Cook, 1979; Gramlich, 1978). Indeed, if the purposes of these programs were training and economic stimulus, neither was a success.

Apparently, little was learned from this experience about the relation between countercyclical policy and national urban policy. In fact, with the U.S. economy in another recession, there is not a firm countercyclical policy.

If business cycles were linked to intergovernmental policy, an essential feature of the system would have to concentrate on more distressed jurisdictions. This raises the thorny problem of identifying those communities most hurt by recession and the severity of the recession in the various regions. The evidence of the past two recessions seems clear—the older manufacturing belt in the Northeast and Midwest was hit hardest (Nelson and Patrick, 1975; Rosen, 1980). Expectations are for a similar regional effect in the next recession (Zamzow, 1980).

An ambivalence—at the federal level—about the "proper" role of state government in state and local government finances may exacerbate some of the problems created by inflation and a slower growing economy (Break, 1980). State governments raised 58% of all state and local government taxes, made 40% of direct expenditures, and accounted for 72% of federal aid in FY 1978. Yet state government is approaching a new crossroads—a redefinition of its fiscal role. The past decade has seen two important, but contradictory, influences on state governmental financing and delivery of services. The first concerns the states' relation to the Federal government and its place in the intergovernmental system. Total grants-in-aid have quadrupled since 1970, but much of this growth has been in direct federal and local grants, with the states being bypassed. In 1978, local governments were directly receiving 28% of total federal aid to state and local governments; in 1970, the figure was 13%. This policy of direct federal-local relations is not inconsistent with the view from some state capitols that city financial emergencies are as much federal as state governmental responsibilities.

Now, as the end of the general revenue sharing authorization approaches, the administration has recommended eliminating the state share. Whether or not state governments have brought this change on themselves by abrogating their responsibility toward urban governments is debatable, but the drift toward reducing the importance of state government in the intergovernmental process is real enough.

There is also a continuing shift of financial responsibility from local to state governments. The state government's share of state and local government taxes rose from 50.7% to 58.5% between 1965 and 1977, and the state's share of direct expenditure increased from 34.9- to 39.9%. The state aid share of total state expenses remained constant between 1965 and 1978, but the state governmental share of health, education, and welfare direct spending increased markedly. States may not have done all that they should to lift the financing burden from the local property tax, and too little may have been done about city and suburb fiscal disparities, but the trend toward more state fiscal responsibility has continued. A combination of local government tax or expenditure limitations, a more elastic state government tax structure, and high rates of inflation could accentuate this trend.

In fact, the increased federal-local aid flow may have slowed the trend of state financial assumption. Before 1975, state aid had behaved as though it were a highly elastic tax, i.e., for every 1% increase in personal income, there was a 1.6% increase in state aid to local governments. That responsiveness fell to 0.96% in 1976 and 0.69% in 1977.

With resources limited, it is imperative to develop a less ambiguous federal position about the role and responsibility of state governments. Is fiscal centralization to be encouraged? And should states—as a prerequisite to federal assistance—be required to deal with the city and suburb disparities problem?

DEFAULT AND EMERGENCY LOANS

Financial emergencies, if not default, lie ahead for many large cities. If it does nothing else, a national urban policy ought to outline the Federal response. Dealing with New York City ad hoc was excusable: There had been little reason to be concerned with municipal default since the depression. In many respects, the New York City crisis of 1975 was a special case.[7] But how many special cases can there be before a policy response must be made? Cleveland and Wayne County have much in common with New York City in the weakness of the underlying economy, as do many of the other cities commonly appearing on the distressed lists.

Two questions are essential in formulating a federal policy for distressed cities. The first involves defining the conditions necessary for Federal intervention, i.e., what avenues must be exhausted before emergency federal subsidy is warranted? The second is what adjustments must the city make as a condition of receiving the aid. Neither question was clearly thought through, and neither is in the administration's urban policy statement.

On the first issue, one might query the state government having a prior responsibility for city financial problems. Should there be an emergency loan to New York City when New York State runs enough of a surplus to cut taxes? Some would argue that the Clevelands and Detroits are primarily the business of the Ohios and the Michigans, and federal bailouts are a last, desperate resort. The view from the statehouse is likely to be quite different. State governments could well argue that a combination of local autonomy, Federal mandates, and direct federal-local aids have taken much of the control of local fiscal excesses out of their hands. Federal actions stimulated the local fiscal and may have created some of the risk of default, hence, the federal government should participate as at least an equal partner in the bailout. The state argument is strong. To require states to shoulder more responsibility for the fiscal problems of their local governments, the federal government must be less ambiguous about the role of state government in the intergovernmental system. If states are to have first claim on filling the financing gap of cities facing financial emergencies, they might reasonably argue for more control over services level mandates and resources passing through to the local level. If cities' financial conditions are to be viewed independently of state government, then a set of criteria for local fiscal actions that must be taken prior to federal intervention should be established. These might include emergency tax levels, program and employment cutbacks, a wage freeze, and perhaps debt rescheduling.

The second issue is how much must local governments alter their fiscal behavior to continue receiving the emergency loan or grant, and how will the fiscal improvements be monitored? The most important question to be resolved is how the federal government will distribute the burden of an austerity program. Employee layoffs and wage freezes will lay much of the burden on public employees, program cutbacks and tax increases on citizens, and bond repayment stretch-outs or moratoriums on bondholders. A federal policy accomodating a bailout in a period of emergency will implicitly or explicitly make such choices.

Another alternative is to make it clear that the federal government will not rescue cities from default, even in the case of the most severe emergencies. Even as a statement of national policy it would be difficult to make this believable with the history of New York City, Lockheed, and Chrysler. But if local and state governments were convinced that a borrower of last resort was not available, their financial practices may become much more conservative and their fiscal strategies more adverse to risk. Whether that would be in the national interest is precisely the sort of question a reasoned national urban policy would address.

STATE AND LOCAL GOVERNMENT POLICY

A national urban policy is essential. State and local government financial problems will materialize in the 1980s, and a reasoned federal response will be imperative. Yet most of the required adjustments will fall to state and local governments, and the majority are neither distressed nor flush.

The fiscal fates of state and local governments will be determined largely by factors outside their control—inflation, the performance of the national economy, and the level and distribution of federal grants. Still, state and local governments have considerable discretionary powers to influence their financial health during this period of adjustment.

The most popular reform is to offer a program for increases in productivity. It is popular because it does not cost the taxpayer, can be used as a basis to reward public employees, and, best of all, its success or failure cannot be measured. The need for, and possibilities of, state and local government employee increases in productivity make great material for discussion, but do not balance budgets. A related issue is whether the tone of the productivity discussion might change with rising materials and energy costs. Heretofore much of the attention had centered on whether capital could somehow be substituted for labor, thereby increasing output and reducing the use of the relatively expensive labor factor. If materials and energy costs continue to rise at present rates relative to labor costs, the enthusiasm for new technologies in the public sector may cool.

A second strategy is the use of tax and subsidy policy to stimulate regional economic development (Schroeder and Blackley, 1979a; Schmenner, 1978; see also Wasylenko, this volume). State and local governments in growing and declining regions attempt to improve their competitiveness as a business location by offering various kinds of subsidies, e.g., tax abatements, tax holidays, subsidized loans, grants of land, and the like. Whether these subsidies have actually contributed to local economic development is as debatable as the issue of whether the induced revenue gains from new business have exceeded the expenditure costs.

Retrenchment—adjusting public service levels and the growth in expenditures to reflect the ability to finance—is probably the most important strategy for governments in declining regions. It involves cuts in service levels and employment, a more realistic look at the kinds of compensation and benefit levels that can be afforded, and a careful conservation of those capital resources available. With the latter, one would expect to see a great deal more emphasis placed on maintenance and renovation of the existing capital stock than on the construction of new capital facilities.[8] The austerity programs in some cities have included these kinds of adjustments, but other public policies have been surprising. Relative tax burdens have gone up in the declining region, the fiscal limitation movement has pretty much been limited to the Sunbelt, and public employment rolls in the declining region have expanded in the past two years.

In the growing regions, local governments also face serious adjustment problems requiring them to carefully plan the growth in their budgets. The problems essentially are how much a government should grow and how fast this growth should occur. The mistakes of governments in the older region might be avoided if the long-term expenditure implications of fiscal decisions are evaluated against the potential long-term growth in the local resource base. Fiscal planning and forecasting is a new art, but is being used effectively in many cities, especially those in the growing region (Bahl and Schroeder, 1979).

The most pressing fiscal adjustment problems are keeping the development of infrastructure in step with population and employment growth. With rising material and capital costs and the prospects for less federal aid, this could become a serious bottleneck to growth. At the same time, there is the danger of allowing growth to become too rapid and uncontrolled, leading fiscal development beyond the possibility of careful, long-term budgetary planning.

STATE AND LOCAL GOVERNMENT FINANCES: THE NEXT FIVE YEARS

The principles of a national urban policy and optimal fiscal adjustments by state and local governments are more wishful thinking than realistic expectations. The likely performance over the next five years will involve a series of financial crises and ad hoc federal responses. The following would not seem an unreasonable scenario:[9]

- The national economy will go through a recession and begin a period of slow real growth. Inflation rates will remain high.
- Some local governments—mostly, but not exclusively, large cities in the North—will either default or be unable to meet their expenditure commitments. A round of public employee layoffs—reminiscent of 1975/1976—will probably take place.
- Despite the recognition of capital obsolescence problems, the quality of the capital stock, especially in the older regions, will continue to deteriorate. Higher interest rates, inflation, reduced federal aid, and pressing financial problems will push state and

local governments to further "defer" capital construction, maintenance, and renovation.
- With rising energy prices, some of the oil- and gas-rich states will experience extraordinary revenue increases and amass considerable surplus funds.
- The next five years will see another catch-up in public employee compensation rates (Grosskopf, this volume). This lagged effect of recent year's deferred compensation increases will be further stimulated by the currently high inflation rate and will account for virtually all of the public expenditure increases of some jurisdictions. The increase in average wages will be especially rapid in the South, where average wages are relatively lower, and where unionization is increasing.
- *Relative* levels of tax burdens will rise in many states in the growing regions in response to increasing costs and service quality and will decline in the Northeast as austerity programs begin to take hold.
- The limitation movement will not significantly slow the rate of state and local government spending after the early 1980s.
- Federal policy toward state and local government finances will remain ad hoc, and there will be no guiding principles. The overall level of federal grants (in real terms) will likely decline and less targeting might be expected during the next five years as the growing region more forcefully makes its point about rural poverty.

These guesses would be altered by either a coherent federal policy toward state and local government finances or by a better performing U.S. economy. In the last analysis, there could be no better national urban policy than a low inflation rate and a strong growth in GNP.

NOTES

1. For state-by state-projections of this slowdown, see Bahl et al. (1979).
2. This may be little more than speculation, since there is no evidence that taxes have a significant effect on the growth of regional employment (Wasylenko, this volume).
3. For a discussion of the possibilities, see Clark and Menefee (forthcoming).
4. Assuming that a faster population growth implies a faster real GNP growth rate.

5. There is no consensus in the literature on consumption about the effects of a changing age distribution on the marginal or average propensity to consume. For a summary, see Russell (1979).

6. See Salins (1979) for a useful discussion of gentrification; and see Allman (1978) for an optimistic view of urban conditions.

7. New York City was unique in its size, the broad range of functions for which it had responsibility, and the excesses in its financial management, particularly its short-term borrowing practices. On the other hand, New York City was not at all unique in its declining economic base, loss of population, rising "dependent" population, and slow-growing tax base. For a discussion of the "uniqueness" of New York during this period, see Bahl et al. (1975).

8. For an example of the results of carefully managing the capital stock in a declining city (Cincinnati), see Humphrey et al. (1979).

9. For another view of the future, see International City Managers Association (1979).

REFERENCES

ALLMAN, T. D. (1978) "The urban crisis leaves town." Harpers (December).
BAHL, R. and L. SCHROEDER (1979) "Forcasting local government budgets." Occasional Paper 38, Metropolitan Studies Program, Maxwell School, Syracuse University.
BAHL, R., M. JOHNSON and L. DeBOER (1979) "The fiscal outlook for state governments." Prepared for Hamilton-Rabinovitz, Inc., October.
BAHL, R., M. JOHNSON, and M. WASYLENKO (1980) "State and local government expenditure determinants: the traditional view and a new approach," pp. 65-120 in R. Bahl, J. Burkhead, and B. Jump, Jr. (eds.) Public Employment and State and Local Government Finance. Cambridge, MA: Ballinger.
BAHL, R., A. CAMPBELL, D. GREYTAK, B. JUMP, Jr., and D. PURYEAR (1975) Impact of Economic Base, Erosion, Inflation, and Retirement Costs on Local Governments. Testimony on fiscal relations in the American federal system. U.S., House, Subcommittee on Government Operations, 94th Cong., 1st sess., 15 July. Washington: Government Printing Office.
BARRO, S. M. (1978) The Urban Impacts of Federal Policies: Volume III, Fiscal Conditions. Santa Monica, CA: Rand.
BREAK, G. (1980) Setting National Priorities: Agenda for the 1980s. Washington, DC: Brookings.
BURKHEAD, J. (1979) "Balance the federal budget" Public Affairs Comment (May). (LBJ School of Public Affairs, University of Texas)
CLARK, R. L. and J. A. MENEFEE (forthcoming) "Economic responses to demographic fluctuations," in U.S., Joint Economic Committee (ed.) Special Study on Economic Change. Washington: Government Printing Office.
COOK, R. (1979) "Fiscal implications of CETA public service employment," pp. 193-228 in K. Hubbell (ed.) Fiscal Crises in American Cities: The Federal Response. Cambridge, MA: Ballinger.
GOODMAN, J. (1979) "The future's poor: projecting the population eligible for federal housing assistance." Socio-Economic Planning Sciences 13: 117-125.
GRAMLICH, F. (1978) "State and local government budgets the day after it rained." Brookings Papers on Economic Activity 1: 191-214.

HUMPHREY, N., PETERSON, G., and P. WILSON (1979) The Future of Cincinnati's Capital Plant. Washington, DC: Urban Institute.
International City Manager's Association (1979) New Worlds of Service. Report to the Professions from the ICMA Committee on Future Horizons. Washington, DC: ICMA.
LADD, H. F. (1979) "An economic evaluation of state limitations on local taxing and spending powers." National Tax J. 31: 1-18.
MULLER, T. (1976) Growing and Declining Urban Areas: A Fiscal Comparison. Washington, DC: Urban Institute.
MUNNEL, A. (1979) Pensions for Public Employees. Washington, DC: National Planning Association.
NELSON, K. and C. PATRICK (1975) Decentralization of Employment During the 1969-1972 Business Cycle: The National and Regional Record. Oak Ridge, TN: Oak Ridge National Laboratory.
N.Y. State Economic Development Board (1978) 1978 Official Household Projections for New York State Counties. Albany: Author.
OLSEN, R. J. (1977) Multiregion: A Simulation-Forecasting Model of BEA Area Population and Employment. Oak Ridge, TN: Oak Ridge National Laboratory.
ROSEN, R. (1980) "Identifying states and areas prone to high and low unemployment." Monthly Labor Rev. 103: 20-24.
RUSSELL, L. (1979) "The macroeconomic effects of changes in the age structure of the population," pp. 23-49 in M. B. Ballabon (ed.) Economic Perspectives: An Annual Survey of Economics, vol. 1. Amsterdam: OPA.
SALINS, P. (1979) "The limits of gentrification." New York Affairs 5: 3-14.
SANDERS, N. C. (1979) "The U.S. economy to 1990: two projections for growth," pp. 12-24 in U.S., Department of Labor, Bureau of Labor Statistics (ed.) Employment Projections for the 1980s, Bulletin 2030. Washington: Government Printing Office.
SCHMENNER, R. (1978) The Manufacturing Location Decision: Evidence from Cincinnati and New England. Cambridge, MA: Harvard Business School and Harvard-MIT Center for Urban Studies.
SCHROEDER, L. and P. BLACKLEY (1979) "State and local government locational Incentive programs and small businesses in region II." Prepared for the Small Business Administration Project, Syracuse University.
SHAPIRO, P., D. PURYEAR, and J. ROSS (1979) "Tax and expenditure limitations in retrospect and prospect." National Tax J. 32: 1-11.
U.S. Congress, Congressional Budget Office (1980) Five-Year Budget Projections: Fiscal Years 1981-1985, A Report to the Senate and House Committees on the Budget: Part II. Washington: Government Printing Office.
U.S. Congress, Joint Economic Committee (1980) The 1980 Joint Economic Report, 28 February. Washington: Government Printing Office.
U.S. Advisory Commission on Intergovernmental Relations (1977) State Limitations on Local Taxes and Expenditures. Washington: Government Printing Office.
U.S. Department of Commerce, Bureau of the Census (1977) Population Projections of the U.S.: 1877-2050. Direct Population Reports, Series P25 No. 704 (July). Washington: Government Printing Office.
U.S. Department of Commerce, Bureau of Economic Analysis, Regional Economic Analysis Division (1977) Population, Personal Income, and Earnings by State: Projections to 2000. Washington: Government Printing Office.
U.S. Department of Energy, Oak Ridge National Laboratory (1978) Long-Term Projections of Population and Employment for Regions of the United States. Oak Ridge, TN: Oak Ridge National Laboratory.

U.S. Department of the Interior, Water Resources Council (1974) 1972 OBERS Projections. Series E population (April). Washington: Government Printing Office.

U.S. Office of the White House Press Secretary (1978) New Partnership to Preserve America's Communities. 27 March. Washington: Government Printing Office.

U.S. Department of Treasury, Comptroller General (1980) An Actuarial and Economic Analysis of State and Local Government Pension Plans, February. Washington: Government Printing Office.

WEINSTEIN, B. (1979) Cost-of-Living Adjustments for Federal Grants in Aid: A Negative View. Research Triangle, NC: Southern Growth Policies Board.

ZAMZOW, J. (1980) The Current Recession: Its Regional Impact. Hearings Before the Joint Economic Committee, 16 October. Washington: Government Printing Office.

8

Fiscal Problems and Issues in Scandinavian Cities

JOERGEN R. LOTZ
Okonomidirektoratet, Copenhagen

□ URBAN PROBLEMS ARE NOT A QUESTION of the absolute size of communities. Scandinavian cities are not large by international standards (see Table 8.1), but they have problems similar to those of large cities in other countries. Scandinavian metropolitan regions suffer from poor quality housing, noise and air pollution, and the like in the center relative to the outer districts. Such disparities affect residential location choices and, hence, relative taxable capacity and expenditure needs of the central cities vs. the outlying jurisdictions. This is a survey of the literature concerning these problems and proposed reforms in the Scandinavian countries. Special attention is paid to the grant systems used to equalize taxable capacity and expenditure needs.

POPULATION STRUCTURE

The migration of families from the central municipalities may be explained in part by the changing stock and quality of housing in the metropolitan region. The quality of housing improved substantially in postwar years, and, as in other countries, a large supply of good quality modern homes became available. But new construction took place where there was

TABLE 8.1 The Population Structure of Scandinavian Capitals, 1978 (amounts in thousands)

Population	Denmark: Copenhagen	Finland: Helsinki	Norway: Oslo	Sweden: Stockholm
Population; nation	5113	4727	4026	8222
central municipality	515	488	460	658
metropolitan region	1755	748	821	1375
Central Municipality as a percentage of:				
metropolitan region	29.3%	65.2%	56.0%	47.1%
nation	10.1	10.3	11.4	7.9

vacant and relatively cheap land, and this was often in the outer parts of the metropolitan area. As a result, the housing stock of the city could no longer meet the demands—especially of families with good and increasing incomes.

Table 8.2 gives the relative distribution of dwellings, by number of rooms, in the central municipalities. It can be seen from the two top rows of Table 8.2 that the housing standard in Sweden and Finland, by size, is lower than in Denmark. Though the quality of central district housing seems poorer in all four countries, Stockholm and Oslo would seem more attractive for Swedes and Norwegians than Copenhagen is for Danes. Such disparities between central and outlying districts explain the exodus of middle and upper income taxpayers from the central cities.

The population disparities accompanying these housing disparities are illustrated in the following two tables. Table 8.3 reveals that the population of the central municipalities is declining in all four countries. Though a stagnation trend may be found in the population of the metropolitan regions (and of the nations), the proportion of the population living in the central municipality is declining.

Comparing the age distributions of the central municipalities with the national age distributions shows a uniform pattern: Cities have a high proportion of old people and a small propor-

TABLE 8.2 Housing Quality Measured by Number of Rooms: Percentage Distribution in 1976

Dwelling Unit Size	Denmark		Finland		Norway		Sweden	
	Copenhagen	Nation	Helsinki	Nation	Oslo	Nation	Stockholm	Nation
1 room*	13.5	7.2	32.5	13.5	22.2	15.0	33.4	16.7
2 room	41.2	18.5	34.9	23.2	26.3	15.0	28.2	24.9
3 room	26.0	25.2	18.2	26.7	27.6	21.0	21.4	25.9
4 room	13.1	25.4	8.9	17.5	15.2	49.0	9.4	17.6
5 or more rooms	6.2	23.1	4.6	15.9	8.7		7.6	14.6
Not known			0.9				0.1	
Total	100.0	100.0	100.0	100.0	100.0	100.0	100.0	100.0

SOURCE: Bypolitikk, Norges Offertlige Utredninger (1979: 5).
*Excluding kitchen.

TABLE 8.3 Population Size and Growth in Scandinavian Capitals, 1960-1978, and Projections for 1985 (in thousands)

Year	Copenhagen City	Region	City as a Percentage of Region	Helsinki City	Region	City as a Percentage of Region
1960	721	1608	44.8	448	546	82.1
1965	679	1678	40.5	495	629	78.7
1970	623	1753	35.5	523	694	75.4
1978	515	1755	29.3	488	748	65.2
1985	460	n.a.	---	474	n.a.	---
	Oslo			Stockholm		
1960	475	709	67.0	805	1163	69.2
1965	485	750	64.7	790	1261	62.6
1970	474	800	59.3	752	1349	55.7
1978	460	821	56.0	658	1375	47.9
1985	434	n.a.	---	611	n.a.	---

tion of children (see Table 8.4). These differences are consistent with the hypothesis that young people choose to live in the central city during their education but move away when they have children and commute to work in the cities. (All four cities have a larger number of jobs than employed residents.) Since it is the older population that stays behind, natural population growth slows distorting the relative age composition even more.

The fiscal implications of these population changes are not completely clear. The loss of middle- and upper-income families depletes taxpaying ability and commuters, and an increasing share of the elderly exert expenditure pressure, since old people need costly hospitals, institutions, and care in their homes. On the other hand, a smaller population with fewer children may reduce expenditure pressures.

LIVING CONDITIONS IN CENTRAL CITIES

Aase and Dale (1978: 52) divided Norwegian municipalities into types, ranging from the most dense and urban to the most

TABLE 8.4 Age Distribution in Scandinavian Capitals and Nations, 1977 (percentage distributions)

	Denmark		Finalnd	
	Copenhagen	Nation	Helsinki	Nation
0-14 years old	12.9	22.3	16.8	21.8
15-64 years old	64.2	64.0	69.8	67.4
65 and over	22.9	13.7	13.4	10.8
TOTAL	100.0	100.0	100.0	100.0
	Norway		Sweden	
	Oslo	Nation	Stockholm	Nation
0-14 years old	15.7	23.4	13.1	20.5
15-64 years old	65.3	62.6	66.6	64.0
65 and over	19.0	14.0	20.3	15.5
TOTAL	100.0	100.0	100.0	100.0

sparsely populated and rural municipalities, and have developed and compared different social indicators for these types of municipalities. They found that income, job satisfaction, and quality of housing and education are highest in more urbanized areas. But for noise, air pollution, and conditions for raising children, larger cities scored low. They also found that contacts with friends and families are more satisfactory in rural municipalities.

This paints a picture of "compensations"—if one type of municipality has advantages in one field, it is likely to be wanting in another. In choosing a residence, people have a choice of the kind of values they prefer.

The same type of analysis carried out for districts within a city shows another pattern: Some districts score low values on all social indicators, while others score high on all counts. There is a "polarization" of living conditions in cities, creating a need for more social assistance than in smaller municipalities. A recent Danish study of poverty confirms the Norwegian findings (Hansen, 1980: 1226). It concluded that when the most deprived citizens were defined by their residency, it was nearly always Copenhagen mentioned as the worst case. It is in particular from these deprived city districts that people move away—if they can. The vacant dwellings are then taken up by those who are not able to find a better place to live. These districts, and their growth, contribute to the present financial squeeze of the Scandinavian cities.

The squeeze is not felt to the same extent everywhere. Stockholm has benefitted from the Swedish local company tax and still maintains a comfortable fiscal position. One might speculate that Oslo may feel the pressures soon, and Copenhagen, whose income level is declining and approaching the national average, would appear to be in the most difficult position.

ADMINISTRATION AND FUNCTION

To evaluate the fiscal situation of a city, it is important to know the functions allocated to local government and the administrative set-up of city governments. Table 8.5 shows how the elected councils vary in size. They have all formed chairmanships and committees with special responsibilities for organizing the work and for supervising different branches of the administration.

For organizing the administration, there are two, quite different models, reflecting different views on the need to separate the legislative and administrative function. For Helsinki and Oslo, a "British" model is chosen, i.e., the administration is run by permanent directors, in particular the powerful directors of finance and six or seven directors for different functions. Copenhagen and Stockholm use a "continental" model, and the chief administrators are elected by the council for a four-year (in Stockholm a three-year) term.

The advantages and disadvantages of the two models are much discussed. The British model ensures more stability and better possibilities for planning, while the continental model gives more political influence and is a more efficient way of decentralizing political power. But permanent directors may gain considerable influence under the continental model, unless—as in Sweden—the entire top administration is replaced when a new party comes to power.

Table 8.5 also shows the functions assigned by the central governments to the city administrations. In the case of Norway and Denmark, some of the functions listed for the central municipalities are in other countries assigned to counties. (Copenhagen and Oslo are exempted from the two-layer structure of local government.) In Stockholm, the county runs hospitals, health, and public transportation, but in Copenhagen, public transportation is placed in a special metropolitan agency.

In sum, Scandinavian municipalities are assigned a range of responsibilities that is quite broad by European standards and about in line with those provided in many American cities. It is particularly important to note that Scandinavian municipalities have considerable responsibility for the redistribution. This underlines the fiscal importance of the outward migration described above and is different from the situation facing American and other European cities. This range of expenditure responsibility emphasizes the need for a national concern with city finances. The social policy carried out by the municipal councils is, in effect, the social policy of the nation. As will be noted below, the success of social policy in these countries depends on the design of the grant systems supporting the finances of Scandinavian cities.

LOCAL TAXATION AND GRANTS

The dependence of the Scandanavian capitals on various forms of financing is shown in Table 8.6. Taxation accounts for between one-third and one-half, while the balance is financed by grants, fees, and charges. The tax and grant items are of particular interest, since these are the amounts to be paid in taxes one way or another. The amount of fees and charges reflects different degrees of "netting" revenues against expenditures, differences in volume of sale of services to other local governments, and so forth.

TABLE 8.5 Administrative Organization and Functions of Scandinavian Capitals

	Denmark: Copenhagen	Finland: Helsinki	Norway: Oslo	Sweden: Stockholm
Number of seats in elected council	55	85	85	101
Administrative leadership;				
elected:	1 + 6			1 + 8
permanent:		1 + 6	1 + 6	
Functions:				
Hospitals	+	+[a]	+	−
Health	+	+	+	−
Secondary schools	+	+	+	+
Social assistance	+	+	+	+
Childrens day-care	+	+	+	+
Old age care	+	+	+	+
Primary schools	+	+	+	+
Libraries, culture	+	+	+	+
Streets, garbage	+	+	+	+
Public utilities	+	+[b]	+	+
Public transportation	−	+[b]	+	−

[a] Organized in municipal ad hoc associations.
[b] Organized in municipal companies.

Central government grants are smaller than own-tax revenues in all cases, ranging from one-fourth in Helsinki to two-thirds in Copenhagen. Grants play a more important role in Danish local finances than in the finances of other Scandinavian countries.

Local government in the Scandinavian countries depends heavily on their systems of *local income taxation.* The tax is on the same income base as used by the central government and is levied on individual incomes as well as on companies, except in Denmark where companies are exempted from local income tax. Local authorities are free to set their own tax rates in Denmark, Finland, and Sweden. There is wide variation in local income tax rates, ranging from 13% to 19% in Finland, 14% to 29% in Denmark, and 23% to 31% in Sweden. Norway introduced a maximum rate limit on the local income tax. Since the 23% maximum rate is levied in all 454 municipalities, this has effectively changed the local income tax into a grant proportional to taxable incomes.

TABLE 8.6 Financing Municipal Expenditures: 1980 Budgeted Amounts

	Copenhagen Danish kroner	per- centage	Helsinki F. Mark	per- centage	Oslo Norw. kroner	per- centage	Stockholm Sw. kroner	per- centage
Tax revenues	6.3	45	2.3	36	4.0	44	4.2	43
Central government grants[a]	4.5[b]	32	0.5	8	1.4	16	1.8	19
Fees and charges, etc.[c]	3.0	21	2.6	40	2.8	31	3.1	32
Net interest	-	-	0.9	14[d]	0.1	1	0.2	2
Net loans and liquidity	0.3	2	0.2	2[d]	0.7	8	0.4	4
Total	14.1[b]	100	6.5	100	9.0	100	9.7	100

[a] Including grants from all other governmental sectors.
[b] Excluding expenditure financed 100% by central government.
[c] Including public utility charges and proceeds from sale of services to other governments (includes rents).
[d] Gross.

In Norway and Denmark, local property taxes are levied, though local councils in Denmark have a choice between the income and property tax. The property tax is in Norway about 5% of local tax revenues and is levied on property values, while in Denmark the revenue is well above 10% of local tax revenues and levied on land alone.

GRANTS POLICIES

Central government grants in the Scandinavian countries are basically of two types. *Conditional grants* are those related to the actual expenditure of the local government in question; e.g., they can be a percentage of the local expenditure on a given function. *General grants* are independent of local expenditure decisions and in the Scandinavian countries are associated with differences in taxable capacity or expenditure needs.

Conditional grants were the dominant grant-form at the beginning of the 1960s. At least four arguments were made in their support (Broman, 1974). First, when the central government asks local government to perform certain functions, it is made responsible for some share of the costs. Moreover, a local share is important to prevent excessive or inefficient spending by local authorities.

Second, the tax burden should be distributed according to the ability to pay; therefore, central government taxation is preferable because of the progressive rate schedule of the central government income tax (compared to the single tax rate of local governments). This advantage became less important after the introduction of the central government value added tax (which is now the "marginal" tax revenue source). Third, variation in local tax rates should be equalized to account for differences in taxable capacity and expenditure needs.

Finally, there is, in some cases, a need for central government to induce local governments to provide certain desired public service standards, and conditional grants are believed to result in higher local expenditure than would otherwise have been decided. This hypothesis of expenditure stimulation has not been empirically tested in the Scandinavian countries. But incentives may be too strong, so that the activity in question is expanded beyond central government intentions. For example, incentives for increased educational spending may reduce the number of children per teacher, even though the central government may have intended more classes. Therefore, conditional grants often depend on certain standards, e.g., a minimum class size.

The first two arguments are valid also for the introduction of general grants that has been under way since the 1960s. The third argument also supports general grants—they are easier to design to compensate for differences in taxable capacity and expenditure needs, while leaving differences in service levels for own tax-financing (Figure 8.1).

While general grants have several advantages over conditional grants, conditional grants are the solution when incentives for certain local expenditure decisions are deemed necessary, or when equalization of differences in expenditure needs is insufficient. Conditional grants still dominate in Norway, although a proposal to introduce a county general grant to replace the specific grant for hospitals was presented in 1979. It was designed along the lines of the Danish model (see below) though without those criteria for social conditions recently introduced in Denmark. In addition, there are compensation payments to municipalities with low taxable incomes or otherwise deemed in need by the central government. These are financed by a surcharge on the local income tax of 2% and by a central government subsidy.

In Sweden, too, most grants to local governments are conditional, but there are two kinds of general grants. One compensates for the local revenue loss because of the personal exemption under the income tax. The grant is, roughly, equal to the personal exemption multiplied by the average municipal tax rate.[1]

The other Swedish general grant—a resource grant—combines low taxable capacity with a crude measure of need. All municipalities are divided into so-called tax-power classes based on geographical criteria. Municipalities in the most northern

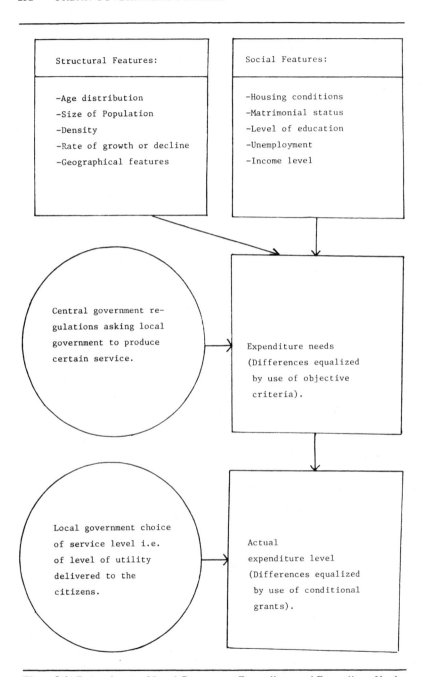

Figure 8.1 Determinants of Local Government Expenditure and Expenditure Needs

regions, where prices are high because of costly transportation and living is more demanding in terms of heating, special equipment, and so forth, are deemed to need 130% of the average tax capacity. The lowest tax power class is considered to need only 95% of the national average tax capacity; Stockholm belongs to the latter. This governmental grant is a deficiency payment calculated as the actual tax rate multiplied with the "equalized" tax base. Municipalities that have more than the minimum necessary tax base do not have to pay contributions, because the costs of the program are carried by the central government. Stockholm does not receive resource grants, because its average income exceeds the tax base deemed necessary.

In Denmark, general grants—"block" grants—were introduced during the 1970s. They are larger than in the other Scandinavian countries, and now acount to 60% of grants to Danish local government. Danish general grants compensate for differences in expenditure needs. The distribution depends on a number of criteria meant to indicate different aspects of local government expenditure needs (Table 8.7). Twenty-five percent of the grants are distributed in proportion to the number of inhabitants, 3% are related to the number of small children, and so forth. The last two criteria, introduced in 1980, are of special interest: They represent the social conditions of the cities resulting from poor housing, polarization, and the like.

Since block grants account for 25% of local government financing, and since they are distributed proportional to expenditure needs, 25% of differences in expenditure needs are equalized through block grant financing. An additional 40% of such differences are equalized in the metropolitan region through a special equalization scheme.

On the resources side there is a higher degree of equalization. Local units with average tax bases below average in the metropolitan region receive compensation for 85% to 100% of the deficiency. Units with tax bases above the average pay up to 70% to 85% of this shortfall to support low-income units. The rest of the subsidy is financed by the central government. The equalization is at about half of this level outside the Copenhagen metropolitan region.

TABLE 8.7 Formula (counties and municipalities) for Distribution of Block Grants: Denmark, 1980

Criteria (nationwide; expenditure needs in parentheses)	[Tranche] Proportion of total grant distributed according to this criterion
– The share of the total population (administration, social welfare)	0.25
– The share of the number of children aged 0-6 years (kindergarten etc.)	0.03
– The share of the number of schoolchildren aged 7-16 years (primary school system)	0.18
– The share of the number of young people of 15-19 years (secondary school system)	0.03
– The share of the number of persons 0-20 years old (certain institutions)	0.01
– The share of the number of persons 17-19 years old (social assistance)	0.01
– The share of the number of people 65-74 years old (homes for the aged, etc.)	0.01
– The share of the number of people more than 75 years old (homes for the aged, etc.)	0.05
– Share of number of apartments for rent, weighted by the rent (rent subsidies)	0.01
– Share of number of "standardized" hospital beds. Assumes that all citizens have the national average sickness for their age group (hospitals)	0.25
– Share of number of kilometers of road (road maintenance)	0.04
– The share of number of children with only one parent (social needs)	0.09
– Share of number of old people living alone (social needs)	0.04
	1.00

NOTE: There are two block grants, each with its own measure of needs. One is distributed to counties and the other to municipalities (Copenhagen receives a share of both).

For the future, efforts to design better grants or equalization systems will dominate the policy on urban government finances in the Scandinavian countries during the 1980s. In Sweden and Finland, the discussion has not yet begun, though the financial squeeze is felt by some Swedish cities. In Norway, a simple model for block grants has just been proposed—without criteria for social needs—to replace some conditional grants. In Denmark, several types of municipalities are eager to see the distribution of block grants altered to reflect even more refined criteria, and a commission to review the whole local government financial structure was formed in 1980. To understand the considerable interest in reforms in this field, it is necessary to understand that social assistance in the cities is synonymous with the social policy of the country. If the central governments wants an effective social assistance policy, it must provide local governments with the means for carrying it out.

EXPENDITURE NEEDS

The immediate question for Scandinavian urban finances is to find a better measure of expenditure needs and to improve the equalization feature of the local government grant system. To illustrate the differences in expenditure needs, Figure 8.1 gives a crude model for determining local government expenditure. The central government assigns local governments the responsibility to perform certain functions, while preserving other functions (e.g., defense) for themselves.

In the cases where central government instructions to local governments are detailed, it may be possible to calculate the expected expenditure for a given local authority by referring to its structure. For example, the number of school-age children may be multiplied by the average cost of meeting the centrally given standards for schools. Even for services that are more or less indivisible and where excluding someone from benefitting from the service is difficult and costly, one may base the calculation of the expected expenditure on the size of the authority as measured by population in general. An estimate of the expenditure necessary to meet central government norms could, in such cases, be suggested by multiplying the population with some centrally determined per-capita "standard expenditure."

This kind of approximation of needs becomes much more difficult when central government instructions are not precise. This would allow local governments to define for themselves the methods and standards for carrying out the assigned functions. Such is the more typical case, at least in the Scandinavian countries. More indirect measures of expenditure needs must be constructed in these cases. The starting point is the available information, such as the size and age distribution of the population of a local authority, its housing conditions, the number of single parents, number of unemployed, its income and the like. These variables have been used in many countries as proxies for conditions causing expenditure variation (Cameron and Lotz, 1981). Still, two local identical governments may choose different expenditure levels as indicated in Figure 8.1. Such variation is the purpose of decentralization, i.e., local authorities can choose their own service levels. But public service variations should be reflected in variations in local tax rates, not as a result of variations in grants received.

One possibility for estimating expenditure needs in these cases is to use actual expenditures as a measure of expenditure needs. The instrument for equalization in this case is conditional grants. But a side-effect of this approach is that those with high service levels receive higher grants, which is contrary to equalization. Therefore, to give the notion of expenditure needs meaning one must think about what level of expenditure the regulation causes the average local authority to decide on, given the structure and social conditions in each case. This suggests a statistical solution, identifying some kind of "average" behavior by local councils. The purpose of applying statistical methods is to relate the criteria mentioned at the top of Figure 8.1 to the actual expenditure decisions at the bottom, hoping that the effects of varying local government preferences for service levels may disappear as "random variation."

One obvious method for estimating expenditure needs this way is the method developed by American scholars in connection with so-called expenditure-determinant studies (Bahl et al., 1980). These techniques have, as a matter of fact, been taken up in Europe to estimate expenditure needs. The method means, in its simplest form estimating an equation linking expenditure to criteria. Let U^n be local government expenditure on function n, N^n the age group of the population expected to be served by function n (or some measure of size in case of indivisible services), Y the average income, and S a social indicator chosen among the "social features" in Figure 8.1. The parameters a_0, a_1 and a_2 can be estimated on data for all local authorities from [1], while ϵ the error term, is supposed to catch differences in service levels:

$$U^n/N^n = a_0 + a_1 S + a_2 Y + \epsilon \qquad [1]$$

Equation 1 will, for any local authority, yield an expected value for U^n, which can be used as a measure of expenditure needs.

The British grant system for England and Wales has, since 1973, been based on such a regression equation, determining local government expenditure as a function of a number of structural and social variables. In Denmark and Norway, regression analysis is used by the cities to serve their special needs.

This approach has been criticized (Cameron and Lotz, 1981). It has been argued that local expenditure variation is not a product of the needs of the people, but of the manipulation of the public employees, in which case it is difficult to give meaning to the concept of expenditure needs. It has also been noted that a proper formulation of a model for local government decision making must be much more refined than suggested by equation 1 (Bahl et al., n.d.). American expenditure-determinant studies now give more attention to the role of developments in local public salaries; but this need not enter the equations when local salaries are centrally negotiated, as is the case in the Scandinavian countries.

A special problem is caused by the effect of resources as measured by the income variable in [1]. If income is positively related to expenditure, it may measure the income effect of demand for social services. (One may also introduce central government grants as an explanatory variable measuring a resource effect.)

Expenditure needs used for equalization purposes should not include the income effect. The British system excludes income from the equation when needs are calculated for specific authorities. In the Danish proposal the resource effect was neutralized by the inclusion of the national average value for resources when needs were to be calculated using [1].[2]

Regression analysis is useful in searching for an index of local government expenditure needs and has demonstrated a tendency to favor the cities.

The reason why cities fare better is that the regression method often replaces systems, defining expenditure needs by simple reference to number of inhabitants or perhaps to the number of people in different age classes or the like. The advantage of the regression method is that it makes it possible to include some proxy measure of social conditions in the measure of needs (Lotz, 1981).

Jackman and Morss (1981) argue that there is a "huge number" of local authority characteristics that may have a bearing on the amount they spend. Since it is impossible to estimate with such a large number of variables (data are not available, and the number of variables may exceed the number of observations), regression analysis is being used instead to estimate what is called an "umbrella" model. The idea is to identify a few characteristics or factors closely associated with a large number of other, true-needs factors. The resulting estimate of need associated with a chosen needs factor measures the spending need directly attributable to that factor and those attributable to other needs-factors that the umbrella variable represents.

For example, the number of "children with only one parent" of an authority may be correlated with local government expenditures per child for kindergarten but may also be correlated to expenditures caused by a large number of other, poorly educated and low-income families living in the same old and cheap neighborhood. The regression analysis does not identify the separate needs of families with one parent—the weight given to this factor may take into account the costs of services needed for them as well as for a number of other families.

The advantages of the umbrella approach is that one can have a measure of needs, though data for the true-needs factor are not available. How needs arise is not well understood—as is often the case in social policy. *The umbrella approach is therefore, in practice, the only way to express social expenditure needs.* This is one reason why a grants policy based on regression results tends to favor cities rather than smaller and rural areas.

The disadvantage is obviously that the "true" model is not known, and that there is a risk that introducing new statistical information, or just the passage of time, can lead to completely different factors being selected in the umbrella model. As already noted, careful examination of results in the United Kingdom and in Denmark suggests that the risk of instability may be exaggerated. In practice, the equations are quite stable if they have included the umbrella variable for social expenditure needs from the beginning (which was not the case in the United Kingdom, where there has been criticism of the consequent fluctuation of the grant formula in years when the social criteria were introduced).

SIZE AND EXPENDITURE NEEDS

A number of studies have shown that there are considerable economies of scale in the production of local government services. Bergstrand and Hederstierna (1980) surveyed a number of cross sectional studies. One found, comparing Swedish port authorities, that there are economies of scale for equipment, but could not find evidence about the costs of employment (Jansson and Ryden, 1979). Another study showed declining costs for larger old-age homes and high schools in Sweden (Tjernstrom, 1972). A third example confirmed the existence of economies of scale for sewage systems in Norway (Rattso, 1979).

Despite theoretical and practical problems of measurement, these studies confirmed that large units of production are associated with low unit costs—at least for some expenditure functions. This finding raises the question of why a number of studies have shown that cities have higher expenditure levels than do smaller, local authorities. Nordstrand (1980) who found such a relationship, thinks that size is proxy for other need-variables, and he showed that size cannot enter the expenditure function for Danish municipalities when tested with a variable measuring social aspects of needs. This suggests that size is related to the polarization of living conditions.

Furthermore, as in other countries, female participation rates in the work force in Scandinavia is higher in the cities than in the rest of the country. This also calls for higher expenditure, because the old and the children then have to be supplied on an institutional basis instead of in the homes, as was the case before women joined the labor force. These and other social factors call for higher levels of local government expenditure in cities than elsewhere.

The costs of bureaucracy may be proportionally higher in large local authorities than in small ones. One Danish study (Mouritzen and Skovsgaard, 1980) suggested that this is true, though data on costs of administration are not very good. (For big authorities, there is a tendency that administration becomes "visible." Administrative reports done by social workers, physicians, or others reported as costs of field work in small municipalities may in big authorities be produced by specialized agencies and classified as "administration," though the work is the same). There may be some extra costs of communication and control for big authorities.

A third explanation for higher costs in cities is that average incomes in the city are often high when compared to the rest of the country. Some of the differences in money income may be only nominal, reflecting different costs of living. But in some cases real incomes may be highest in the city. If this is so, and if demand for local government services is income elastic, there is a resource-effect resulting in higher local government expenditure.[3]

Empirical evidence suggest that these factors outweigh the effects of economies of scale in producing local government services, and the net result is typically higher expenditures per capita in the cities than in the small local authorities. This also seems to be the case if the resource-effect is identified and neutralized, and the conclusion is suggested that expenditure needs are highest for big authorities (Kommunale udgiftsbehov, 1978).

CONCLUSIONS

Financial problems of the cities may be caused by insufficient tax capacity or by large local government expenditure needs. The risk that both conditions are met is greater, the smaller part of the metropolitan region is defined as a central municipality. This is because the housing stock in the center is often old, worn out and cheap, offering shelter for those who are not able to pay for the better and newer homes in the outskirts of the city.

Living condition are often polarized in the cities, and there are groups who are in extreme need of public assistance. Since decentralization of social policy is carried relatively far in the Scandinavian countries, these problems are felt in city finances. Because social policy is a national concern, equalization and grants are important issues to both the central and local governments.

This does not mean that local government taxation is not well developed. Municipal income taxation is a powerful source of finance in Scandinavian countries, but it is only sufficient if taxable capacity is of a size sufficient for the expenditure needs. This is not typically the case for local authorities.

These problems are partly solved by equalization schemes and by central government grants. The traditional grant form is conditional grants related to actual expenditure, but in recent years block-grants have been introduced in a form designed to compensate for differences in measures of taxable capacity and expenditure needs. In Denmark, the measure of expenditure needs is developed to take account of differences in social needs.

There has been much discussion recently on how best to measure local government expenditure needs, and the use of expenditure determinant studies—the basis for distributing block grants in England and Wales—is being developed. The discussion is very much centered on the role of resources as a determinant of expenditure needs, i.e., on how to identify and handle the income effect in assessing expenditure needs for equalization.

Economies of scale in the production of public services are often said to favor cities, and several studies seem to confirm this. But factors such as the social problems of the centers, the bureaucracy necessary for large units, and perhaps the resource effects of high, real incomes, seem to more than outweigh possible economies of scale. Expenditures have empirically been shown to be highest in the larger municipalities, suggesting that this also holds true for expenditure needs.

NOTES

1. This revenue source is considered a tax in Swedish local government budgets.
2. See *Kommunale udgiftsbehov*, Betaenkning nr. 855, November 1978. The proposal to introduce the regression method in Denmark has not been accepted by the government.
3. This last argument is often used as an argument against better equalization of differences in expenditure needs, and this is one reason why neutralization of the income effect, when measuring expenditure needs, may be important as a political condition for better equalization.

REFERENCES

AASE, A. and B. DALE (1978) "Levekar i storby." Norges Offertlige Utredninger 52.
BAHL, R., M. JOHNSON, and M. WASYLENKO (n.d.) "State and local government expenditure determinants: the traditional view and a new approach," in R. Bahl, J. Burkhead, and B. Jump (eds.) Public Employment and State and Local Government Finance. Cambridge, MA: Ballinger.
BERGSTRAND, J. and A. HEDERSTIERNA (1980) "Anvanding av kostnadssamband vid kommunalt beslutsfattande." Presented to the Conference on Nordic Urban Financial Problems, Copenhagen, February.
BROMAN, L. (1974) Kommunal ekonomi, hermodskonsult. Stockholm: Hermodskonsult.
CAMERON, G. and J. LOTZ (1981) "Expenditure needs," in Copenhagen Workshop on Measuring Local Government Needs. Paris: OECD.

HANSEN, E. J. (1980) Fordeling of levevilkarene, vol. 4. Copenhagen: Socialforskningsinstitutet.
JACKMAN, R. and E. MORSS (1981) In G. Cameron and J. Lotz, Copenhagen Workshop on Measuring Local Government Needs. Paris: OECD.
JANSSON, T. O. and RYDEN, I. (1979) Samhalls-ekonomisk for hamnar. Stockholm: EFI.
LOTZ, J. (1981) In G. Cameron and J. Lotz, Copenhagen Workshop on Measuring Local Government Needs. Paris: OECD.
MOURITZEN, P. E. and C. J. SKOVSGAARD (1980) "On the importance of the urban hierarchy for local government service and expenditure levels." Presented to the ECPR Workshop on Urban Policy and Expenditure Patterns, Florence, Italy, March 25-30.
NORDSTRAND, R. (1980) "Variation i primaerkommunale udgifter." Presented to the Conference on Nordic Urban Financial Problems, February.
RATTSO, J. (1979) "Kommunal forurensningspolitik arbejdsrapport." Oslo: Norsk Institut for by- og regionsforskning.
TJERNSTROM, S. (1972) "Kommunale kostnadsstudier." Stockholm, EFI.

9

Urban Finances in Developing Countries

JOHANNES F. LINN
World Bank

☐ URBAN GOVERNMENTS IN THE DEVELOPING countries are commonly faced with problems of urban management that far surpass those experienced by their counterparts in the industrialized countries. Slums and squatter settlements provide shelter for anywhere between 30% and 90% of their cities' populations. Rapid population growth leads to growing urban services deficits, unless major efforts are made in extending services to the large, unserviced sections of the cities. The prevalence of poverty and generally low incomes makes it difficult at every level of government to raise fiscal resources for the provision of urban services at an increased pace. Local authorities are hardest hit since they generally have to make do with the weakest revenue instruments left to their discretion by higher-level authorities. Available resources are often used ineffectively because skilled management and manpower are scarce and are even scarcer for local authorities, since they can make only residual claims on skills available to governments in the developing countries (Beier et al., 1976; Renaud, 1979; Linn, 1979).

AUTHOR'S NOTE: *The author wishes to thank Roy W. Bahl and Douglas H. Keare for their advice and support in carrying out the research on which this chapter is based. The views expressed are those of the author and are not necessarily those of the World Bank.*

The purpose of this chapter is to focus on the urban finance aspects of the difficulties encountered by urban governments in developing countries in trying to cope with rapid urbanization at low levels of income. The analysis deals particularly with the experience in the larger cities, although many of the issues discussed apply equally in the smaller urban areas of developing countries.

The chapter is divided into two parts: The first provides a statistical overview of some dimensions of the urban finance systems in developing countries. It discusses the importance of local government and surveys the diversity of public service responsibilities found among urban governments in developing countries. It then reviews the expenditure and revenue trends of local governments in selected cities and concludes by discussing the limited revenue authority generally faced by urban governments. The second part of the chapter turns to an eclectic overview of some of the major issues in urban finance in developing countries, focusing on the efficiency of urban government, the prevalence and meaning of fiscal gaps, the issues of centralization and fiscal fragmentation, and the scope for urban fiscal reform.

The data on which the analysis in this chapter is based are mainly drawn from ten case studies initiated and supervised by Roy W. Bahl and the author in the context of a comprehensive World Bank study of urban finances in developing countries. These case studies were carried out according to comparable standards of data compilation and analysis based on field visits that permitted access to the financial accounts of city governments in Ahmedabad and Bombay (India), Bogota, Cali, and Cartagena (Colombia), Kingston (Jamaica), Tunis (Tunisia), Jakarta (Indonesia), Manila (The Philippines), and Seoul (Korea). Somewhat less comparable data were drawn from additional case studies of the finances of particular cities carried out under other auspices. In all cases, however, an effort was made to ensure that a minimum of comparability existed in coverage of local government agencies and definition of revenue and expenditure items.

The resulting coverage of cities for the purposes of statistical analysis is less comprehensive and representative than one might ideally wish. If nothing less, the study has established the need for more universally comparable financial reporting. But the major patterns and trends indicate the great diversity as well as some common features in the urban finance systems of the developing countries.

EXPENDITURE AND REVENUE STRUCTURE OF URBAN GOVERNMENT IN DEVELOPING COUNTRIES

THE IMPORTANCE OF URBAN GOVERNMENT

Studies of public finances in developing countries have focused on the activities of national governments but have neglected the role played by subnational public authorities.[1] As a starting point in a discussion of urban finances in developing countries, it is therefore appropriate to consider the importance of subnational and local government.[2]

The concern is not with the degree of decentralization of governmental authority in terms of local autonomy and independence. Rather, the focus on the expenditures of local government reflects concern with the responsibility of local governments for providing public services to the rapidly growing populations.

Accurate data on local authorities in cities are scant; for cross-country comparisons it is useful, as a first approximation, to consider the shares of all subnational levels of government combined. Drawing on a sample of 45 developing countries, Kee (1976) found that in 1970 the average share of subnational government expenditure in total public sector spending was 24%. For a sample of industrialized countries, Kee estimated an average share of 47%.[3] To the extent that these figures are reliable, they indicate that local government in developing countries is of considerably less importance than in the industrialized nations. But when countries are grouped into three categories according to whether they had a high, medium, or low subnational government expenditure share, a different picture emerges from Kee's data. For the sixteen developing countries with a large subnational government sector—eleven of which are in Latin America—the average share of subnational government

TABLE 9.1. Distribution and Growth of Estimated Public Sector Expenditure in Selected Cities by Level of Government

City	Year(s)	Per Capita Local Expenditure (in US$)[a]	Percentage Share in Total Government Expenditure			Percentage Growth Rate in Per Capita Government Expenditure[b]			Source[c]
			Local	State	Central	Local	State	Central	
Colombia									
Bogota	1970-72	59.5	49.9	-	50.1	20.2	-	21.9	Linn (1980a)
Cali	1975	51.4	48.8	6.7	44.5	n.a.	n.a.	n.a.	Linn (1980a)
Cartagena	1969-72	20.0	23.0	8.5	68.4	13.6	15.9	21.9	Linn (1980a)
India									
Bombay	1962/3-1969/70	22.0	41.7	31.5	26.9	6.1	12.7	11.5	Bougeon-Maassen (1976)
Ahmedabad	1965-71	19.7	41.5	49.8	8.7	14.6	9.5	12.5	Bahl (1975)
Mexico									
Mexico City	1966	32.3	18.0	-	82.0	n.a.	n.a.	n.a.	Fried (1972)
Korea									
Seoul	1965-71	31.4	36.3	-	63.7	31.5	-	23.0	Bahl and Wasylenko (1976)
Daegu	1976	41.0	23.0	-	77.0	n.a.	-	n.a.	Smith and Kim (1979)
Gwangju	1976	37.8	21.6	-	78.4	n.a.	-	n.a.	Smith and Kim (1979)
Daejeon	1976	38.4	21.9	-	78.1	n.a.	-	n.a.	Smith and Kim (1979)
Jeonju	1975	31.0	23.5	-	76.5	n.a.	-	n.a.	Smith and Kim (1979)
Nicaragua[d]									
Managua	1972	14.9	15.2	-	84.8	n.a.	-	n.a.	Lacayo et al. (1976)

TABLE 9.1 (continued)

City	Year(s)	Per Capita Local Expenditure (in US$)[a]	Percentage Share in Total Government Expenditure			Percentage Growth Rate in Per Capita Government Expenditure[b]			Source[c]
			Local	State	Central	Local	State	Central	
Philippines Manila	1960-70	7.5	30.5	-	69.5	6.5	-	9.7	Bahl et al. (1976)
Indonesia Jakarta	1972/3	8.3	36.9	-	63.1	n.a.	n.a.	n.a.	Linn et al. (1976)
Tunisia Tunis	1965-70	17.6	17.0	-	83.0	-3.5	-	7.1	Prud'Homme (1975)
Jamaica Kingston	1967/8-71/2	20.7	19.4	-	80.6	13.7	-	18.2	Bougeon-Maassen and Linn (1975)
Iran Tehran	1974	26.2	3.9	-	96.1	n.a.	n.a.	n.a.	Smith (n.d.)

[a] Consolidated expenditure by all local government agencies in the metropolitan area, including (semi-) autonomous local public service enterprise, converted to US$ at the prevailing official exchange rates as shown in IMF, *International Financial Statistics*, May 1976. The figures are for the terminal year of the period shown in the preceding column.
[b] Average annual compound growth rate of per capita expenditures expressed in US$ at the prevailing official exchange rates.
[c] The sources shown indicate the origins of the city (and state) expenditure data. Central government expenditures are taken from the most recent World Bank Economic Report for each country, and include national public enterprises.
[d] Local revenues are used for this city to approximate the level of local expenditures.

spending in total government spending was 46%, quite comparable to the experience in the developed countries. The average share is only 6% for those developing countries where subnational government is relatively unimportant. African countries predominate in this group.

A similarly varied picture emerges in the growth or decline in subnational government relative to the total public sector. Smith (1974) and Kee (1976) have shown that in some developing countries (Brazil, Korea, Sri Lanka, Tanzania, Costa Rica, Greece, Guatemala and Portugal) the importance of subnational government measured by its expenditure share appears to have increased, while it decreased in others (Ecuador, Honduras, Peru, Zambia, Kenya, and Turkey). For yet others, the evidence is mixed (e.g., Colombia and Philippines); and in Korea, it appears that the importance of local government was on the rise during the 1960s (Smith, 1974), but declined thereafter (Smith and Kim, 1979).

A similar picture emerges in local government spending in particular cities. Table 9.1 shows, for a selected number of cities in developing countries, estimates of the share of national, state, and local government expenditures in urban areas.[4] Urban government was very important in Bogota, Cali, Bombay, Ahmedabad, and Jakarta, but of negligible importance in Tehran, Kingston, and Tunis. Grouping cities according to whether local government provides an important share of total public spending follows quite closely the earlier grouping of countries according to the importance of subnational government. This is true even for India, where state governments have substantial expenditure authority.

The growth rates in the spending of the different government levels are also found to vary widely across cities (Table 9.1). In some cities (Ahmedabad and Seoul) local spending increased more rapidly than national government expenditure, while the reverse was true for others (Bogota, Cartagena, Bombay, Manila, and Kingston). Combining countrywide and city-specific evidence, one therefore cannot conclude—as some analysts have done (e.g., Walsh, 1969)—that local government in developing countries has generally lost in importance in recent years. This is true only for some countries and cities. Furthermore, as the

Korean example indicates, ups and downs may occur in the share of local government in total public spending for particular cities or countries over extended periods.

A somewhat clearer pattern emerges when one considers the relative importance of local authorities within countries. Smith (1974) has shown that per-capita spending by local government tends to be higher in the major city than in the country. This is confirmed by the figures in Table 9.1. For those countries where data are available for more than one city, the larger cities show larger per-capita expenditure by local authorities, and local government provides a greater share of total public spending. Further underscoring the importance of local governments in large cities is the fact that the local authorities are frequently the largest *single* agency operating in the metropolitan area, although the various ministries and other agencies of the national and state governments, when combined, may have a larger share of total public expenditure in that city.

In sum, the preceding comparative analysis indicates that urban governments play an important role in providing urban services in many of the large cities in developing countries, and that local government may have an important effect on the development of these cities. In any case, the fiscal and administrative problems of large cities in developing countries deserve more attention than they usually are accorded under the mistaken belief that local government plays only a negligible role.

PUBLIC SERVICE RESPONSIBILITIES OF URBAN GOVERNMENT
IN DEVELOPING COUNTRIES

All urban governments in developing countries provide at least a minimal range of basic urban services commonly attributed to local government throughout the world, i.e. markets, abattoirs, fire protection, street cleaning and lighting, garbage collection, cemeteries, libraries, and some public health facilities. Beyond these common functions, local government responsibilities vary tremendously. Many local governments have full or partial responsibility for providing physical infrastructure, in particular the construction and maintenance of streets, potable water supply, sewer systems, and drainage works.

Telephone and electricity services are typically the responsibility of higher level government agencies, except in a few large cities, especially in Colombia. Primary education is frequently under local government control, while other social services, such as public health and welfare, are rarely major local functions and are often not provided on a significant scale by any governmental agency. Local housing programs are important in a few cities, particularly the former British colonies (e.g., Zambia and Kenya). But generally local public housing programs are small relative to national or state housing programs and frequently cater only to municipal employees. Urban mass transportation is often left to private sector management, operating under the supervision of local or national authorities. But in some cities, such as Bombay, Seoul, and Casablanca, the local authorities are heavily involved. Police protection is mostly a responsibility of national authorities. This large variation in local government responsibilities across developing countries makes a comparison with industrialized countries not very fruitful, but one might mention that urban mass transportation and welfare generally tend to be provided on a large scale and with greater local government involvement in the industrialized nations.

Within developing countries, local governments in the larger cities have greater responsibilities than their counterparts in the smaller cities. One of the reasons for this phenomenon is that the largest cities, and in particular the capital cities, often have a special administrative status combining local and state functions at the level of the metropolitan government and therefore have a greater range of local government responsibilities (e.g., Bogota, Mexico City, Jakarta, and Seoul). This also explains in part why local government in larger cities generally have larger per-capita expenditures and a larger share in total public sector spending than in the smaller cities within a particular country.

One of the most striking phenomena observable throughout the cities of developing countries is the extensive overlap in the responsibilities of local, state, and national authorities. Typically, all levels of government are involved in a particular service. For example, various national, state, and local agencies in Cali, Colombia provide public housing, health services, and education (Bird, 1975); in Jakarta, the national government and the city government share, to varying degrees, in providing water, public health, education, and transportation (Linn et al., 1976). Higher-level governments are often involved in the planning, development, and investment phase of urban service provision by controlling the borrowing of local agencies, by providing capital grants, or by assuming direct responsibility for the investment programs. Other means of central control are special metropolitan development agencies or autonomous service companies in which the higher levels of government take a more active role in controlling the activities of these agencies than is the case with regular local government operations. Good examples are the metropolitan development authorities in Calcutta, Karachi, and Manila, all of which have substantial involvement of higher level government. Another form of overlapping responsibility occurs where national or state-appointed civil servants carry out local government functions, sometimes simultaneously with other national (or state) functions. This is the case in Jakarta, Cali, and in many former French colonies (Linn et al., 1976; Bird, 1975; Walsh, 1969).

Cities in developing countries are, however, not alone in this pattern of overlapping responsibilities of local and higher levels of government. In the industrialized nations, a similar pattern exists (U.S., HUD, 1973; Ehrlicher and Hagemann, 1976) and calls for clearer definitions of expenditure responsibilities are heard in virtually every industrialized country (Marshall, 1969; Walsh, 1969; Apel, 1977).

URBAN EXPENDITURE TRENDS AND PATTERNS

Ideally, it would be interesting to compare the provision of urban services across cities and countries by analyzing the growth and distribution of all public expenditure in cities of developing countries. This could indicate how far the provision of urban services has managed to keep in line with urban population growth, and to what extent priorities in service provision differ between cities and countries. However, the data commonly available do not permit such an analysis on a cross-country basis. For most cities, even a detailed case study approach does not usually result in reliable estimates of expenditure distribution and growth by all governmental agencies. The

TABLE 9.2 Annual Growth Rate and Composition of Expenditure by Local Governments in Selected Cities (in percentages)

City	Years	Population Growth	Total Local Expenditure		Recurrent Local Expenditure		Source
			Growth Rate in Current Prices	Growth Rate in Constant Prices	Growth Rate in Current Prices	Share in Total Local Expenditure[a]	
Colombia							
Bogota	1963-72	6.6	20.5	9.0	21.8	59.0	World Bank Estimates
Cali	1964-74	4.4	21.3	7.9	22.9	72.8	World Bank Estimates
Cartagena	1970-72	5.0	31.0	18.2	31.9	76.4	Linn (1975)
India							
Ahmedabad	1965-71	3.3	9.4	3.7	12.2	88.0	Bahl (1975)
Bombay	1962/3-71/2	3.7	10.5	4.3	11.2	83.7	Bougeon-Maassen (1976)
Madras Corporation	1971/2-75/6	3.7	4.3	-8.8	9.2	74.0	World Bank Estimates
Pakistan							
Karachi	1971/2-74/5	5.6	30.2	7.7	26.7	64.2	World Bank Estimates
Gujranwala	1970/1-74/5	5.8	34.1	16.9	37.6	58.3	Goldfinger (1975)
Indonesia							
Jakarta	1969/70-72/3	4.6	17.8	9.2	19.4	49.3	Linn et al. (1976)
Philippines							
Manila	1960-70	4.9	11.4	6.1	n.a.	n.a.	Bahl et al. (1976)

TABLE 9.2 (continued)

City	Years	Population Growth	Total Local Expenditure		Recurrent Local Expenditure		Source
			Growth Rate in Current Prices	Growth Rate in Constant Prices	Growth Rate in Current Prices	Share in Total Local Expenditure[a]	
Korea							
Seoul	1963-72	7.6	34.5	20.8	22.5	34.7	Bahl and Wasylenko (1976)
Turkey							
Istanbul	1960-70	3.9	5.3	-0.6	5.9	86.8	World Bank Estimates
Tunisia							
Tunis	1966-72	4.0	-3.2	-6.5	0.0	84.8	Prud'Homme (1975)
Kenya							
Nairobi	1960-76	7.0	17.4	4.6	14.2	70.3	World Bank Estimates
Zambia							
Lusaka	1966-72	11.3	14.6	5.7	16.8	84.0	World Bank Estimates
Jamaica							
Kingston	1968/9-72/3	2.8	15.3	8.2	11.1[b]	86.5	Bougeon-Maassen and Linn (1975)
Brazil							
Rio de Janeiro (State of Guabara)[c]	1967-1969[d]	2.7	46.7	16.6	n.a.	71.0	Richardson (1973)

[a] For terminal year of the period under consideration in each city.
[b] 1966/67-1972/73.
[c] Not including autonomous agencies.
[d] 1969 figures are budgeted expenditures.

problem is that higher-level authorities do not generally keep their accounts on a regional or city basis, making it virtually impossible to determine the distribution of their expenditure at the city level.

This study therefore concentrates on the expenditures by local authorities. But an effort was made to provide at least a comprehensive view of the local government sector by including all local public agencies.[5] The analysis of consolidated local government expenditures permits a judgment about the growth (or decline) in the contribution of the local public sector to the provision of urban services and a comparison of expenditure patterns, given the knowledge that different local governments have different responsibilities assigned to them by the higher levels of government.

Table 9.2 shows the growth performance of consolidated local government expenditures in selected cities in developing countries. The table reveals a wide variation in the growth of expenditure in current and constant prices. Some city governments have experienced very rapid expansions in real expenditure, substantially above the rate of increase in the city's population (e.g., Seoul, Cartagena, and Gujranwala). Other cities saw a decline in real spending (Tunis, Istanbul, and Madras), while yet others experienced expenditure growth in real terms, but at rates below population growth, and therefore faced a decline in real per-capita expenditure.

Perhaps the most notable feature of Table 9.2 is that in many of the cities shown, real per-capita expenditures increased over the periods under consideration. Assuming one can infer from this that many local governments succeed in bringing a growing amount of services to urban inhabitants, despite rapid increase in population, a limited resource base, and the frequent constraints placed upon them by higher governmental authorities, this evidence reflects a remarkable achievement.

There are, however, some words of caution. First, expenditure needs may have risen more rapidly than population during the periods under consideration. Therefore, expenditures may not have kept in line with rapidly rising needs, even where they grew more rapidly than population. Second, the rise in real per capita expenditures in many cities may not have been sufficient to meet existing service deficits or to raise the quality of services in areas where they are below the acceptable level. Third, as Table 9.2 also shows, in most cities recurrent expenditures grew more rapidly than total expenditures. While these figures do not necessarily reflect longer term trends, given the lumpiness and high variability of urban investment expenditure from year to year, they may reflect two developments: First, investment functions are increasingly taken over by higher levels of governments; second, local governments are not expanding urban services as rapidly as total local expenditures would seem to indicate, simply because they find it necessary to spend increasing proportions of their available resources in maintaining existing services, rather than expanding coverage of services throughout the growing city. Again, there is a wide variation among cities, but it appears from an inspection of Table 9.2 that those cities that have shown the least buoyant expenditure also tended to have the highest proportion of spending going into current expenditures (Istanbul, Madras, Ahmedabad, Bombay, Tunis, and Nairobi). In contrast, those cities with more buoyant expenditures have managed to spend relatively more of their resources on investment (e.g., Cartagena, Jakarta, Seoul, and Gujranwala).[6]

Data collected on expenditure patterns in 17 cities in eight developing countries show wide variations in the distribution of services between different service categories.[7] This is, of course, only to be expected because of the wide variation in urban government responsibilities, in the relative severity of existing service deficits, and in policy objectives. For example, Colombian cities spent a large share on public utilities not only because of the high degree of responsibility that local governments in Colombia have for these services, but also because of the high priority placed on providing these services in recent years. Ahmedabad emphasized its subsidized milk scheme as a matter of policy choice, and both Bombay and Ahmedabad, again as a matter of policy choice, have had a relatively heavy public involvement in urban mass transportation. Seoul was notable for the heavy discretionary spending on education, reflecting the strong emphasis on educational advancement in Korea. Jakarta and Cartagena were burdened with relatively

TABLE 9.3 Financing of Local Public Expenditures in Selected Cities: Percentage Distribution by Type of Revenue

City	Year	Total Local Revenue	Locally Raised Revenue			Revenue from External Sources			Total	Sources
			Local Taxes	Self-financing Services	Other Local Revenue	Total External Revenue	Grants and Shared Taxes	Net Borrowing[a]		
Francistown (Botswana)	1972	102.9	46.8	56.1	–	-2.9	1.9	-4.8	100.0	World Bank Estimates
Mexico City (Mexico)	1968	101.9	70.9	5.2	25.8	-1.9	8.9	-10.8	100.0	Fried (1972)
La Paz (Bolivia)	1975	97.0	61.9	3.6	31.5	3.0	9.0	-6.0	100.0	World Bank Estimates
Tunis (Tunisia)	1972	93.9	36.8	7.1	50.0	6.1	0.7	5.4	100.0	Prud'Homme (1975)
Kitwe (Zambia)	1975	92.7	35.0	53.1	4.6	7.3	2.2	5.1	100.0	World Bank Estimates
Valencia (Venezuela)	1968	90.8	44.8	13.4	32.6	9.2	9.2	–	100.0	Cannon et. al. (1973)
Lumbumbashi (Zaire)	1972	90.5	72.8	–	17.7	9.5	9.5	–	100.0	World Bank Estimates
Rio de Janeiro (Brazil)[b]	1967	88.4	74.5	7.2	6.7	11.6	1.7	9.9	100.0	Richardson (1973)
Amedabad (India)	1970/1	86.3	38.6	41.8	5.9	13.7	4.2	9.5	100.0	Bahl (1975)
Bombay (India)	1970/1	84.6	37.9	38.7	8.0	15.4	1.0	14.4	100.0	Bougeon-Maassen (1976)
Karachi (Pakistan)	1974/5	84.1	67.6	2.2	14.3	15.9	2.8	13.1	100.0	Kee (1975)
Seoul (Korea)	1971	80.0	30.3	36.3	13.4	19.9	15.8	4.1	100.0	Bahl and Wasylenko (1976)
Jakarta (Indonesia)[c]	1972/3	78.8 (81.9)	40.6 (43.7)	15.2 (15.2)	23.0 (23.0)	21.1 (18.1)	21.1 (18.1)	–	100.0 (100.0)	Linn et al.
Lusaka (Zambia)	1972	78.2	39.3	36.9	2.0	21.8	6.0	15.8	100.0	World Bank Estimates

TABLE 9.3 (continued)

City	Year	Total Local Revenue	Locally Raised Revenue			Revenue from External Sources			Total	Sources
			Local Taxes	Self-financing Services	Other Local Revenue	Total External Revenue	Grants and Shared Taxes	Net Borrowing[a]		
Cali (Colombia)	1974	74.4	15.6	57.5	1.3	25.7	2.8	22.9	100.0	World Bank Estimates
Calcutta Corp. (India)	1974/5	73.8	64.4	—	9.4	26.2	19.4	6.8	100.0	World Bank Estimates
Cartagena (Colombia)	1972	70.4	23.3	43.3	3.8	29.6	12.8	16.8	100.0	Linn (1975)
Mbuji-Mayi (Zaire)	1971	70.2	66.5	—	2.7	29.8	29.8	—	100.0	World Bank Estimates
Manila (Philippines)[d]	1970	70.0	55.0	10.0	5.0	30.0	30.0	—	100.0	Bahl et al.
Bukaru (Zaire)	1971	69.9	67.4	—	2.5	30.1	30.1	—	100.0	World Bank Estimates
Madras (India)	1975/6	69.2	54.5	3.7	11.0	30.8	25.1	5.7	100.0	World Bank Estimates
Bogota (Colombia)[c]	1972	62.5 (72.4)	13.7 (23.6)	48.5 (48.5)	0.3 (0.3)	37.5 (27.6)	14.0 (4.1)	23.5 (23.5)	100.0 (100.0)	World Bank Estimates
Tehran (Iran)	1974	46.9	42.8	—	4.1	53.1	45.2	7.9	100.0	Smith (n.d.)
Kingston (Jamaica)	1971/2	30.1	23.9	2.7	3.4	69.9	67.2	2.7	100.0	Bougeon-Maassen and Linn (1975)
Kinshasa (Zaire)	1971	26.9	25.4	—	1.5	73.1	73.1	—	100.0	World Bank Estimates
Median (Average)		78.8 (76.6)	42.8 (46.0)	7.2 (19.3)	6.7 (11.2)	21.1 (23.4)	9.5 (17.7)	5.1 (5.7)	100.0 (100.0)	

[a] Net borrowing consists of loan financing minus net changes in financial assets or reserves.
[b] Due to exclusion of autonomous agencies the contribution of self-financing service revenues and that of all locally raised revenues are probably understated, while the remaining entries are overstated.
[c] Figures not in brackets include shared taxes under grants; figures in brackets include shared taxes under local taxes.
[d] Total revenues are used instead of total expenditures.

high administrative overheads, while Kingston chose to spend a relatively large share on direct welfare-support services, which is an unusual practice among urban governments in developing countries.

REVENUE STRUCTURE OF URBAN GOVERNMENTS

The preceding sections have shown that the local governments in cities of developing countries often face heavy responsibilities in providing essential services to their rapidly growing urban populations. The purpose of this section is to describe urban government financing of these expenditures.

As a first step it is useful to distinguish between five major types of revenues, three of them from local sources, the other two from external sources. The local sources are: first, locally raised taxes; second, revenues raised from user and other benefit charges (all of these are referred to here as revenues from "self-financing services"); and third, a residual group of locally raised revenues, such as license fees, penalties, and stamp duties. The two external sources of local finance are borrowing and grants from higher-level governments (including shared taxes). The distinction between locally raised and external revenues is useful because there is a presumption that local authorities tend to have more discretion in managing their local sources of finance than is the case for external revenues. But external revenues do not necessarily represent subsidies to the urban economy from outside. Loan finances may be generated from savings within the urban sector and may be fully repaid over time. Higher-level government grants may ultimately draw on national tax revenues generated mainly in the urban sector. This is the case particularly with shared taxes, which are collected by the national government in the city and then entirely or partially remitted to local authorities.

Table 9.3 shows the distribution of revenues according to these financing sources in 17 cities of developing countries, ordered according to the importance of locally raised revenues in financing total local government expenditures. The share of locally raised revenues in financing total expenditures ranged from over 100% in Francistown and Mexico City,[8] to an exceptionally low share of 30% in Kingston. More typically, however, between 60% and 90% of local expenditures were financed from local sources, with a median of 79%.

Local taxes on average provided more than half of locally raised revenues, while self-financing service revenues contributed 25% and other local revenues the remainder. Between cities, a wide variety of local financing patterns prevails. The Colombian cities are notable for their relatively limited reliance on local taxes and a much heavier emphasis on self-financing service revenues. This is related to the fact that in the large Colombian cities, local government tends to provide the major urban public utilities (water, sewerage, electricity, and telephones); but it also reflects the relatively heavy Colombian emphasis on benefit-related charges in financing urban infrastructure.

Mexico City, Karachi, Jakarta, Manila, Madras, and Bukaru relied heavily on local taxes. This, in part, results from the fact that in those cities local government is responsible for services not readily lending themselves to self-financing, such as education, public health, fire protection, markets, parks, and recreation. But it also reflects the fact that at least in some of these cities relatively little attention was given in the past to financing services through user charges. For instance, in many of the former British colonies, water supply is financed from general property taxes; and even where benefit charging systems exist, such as in Jakarta and Manila, they rarely contribute substantial amounts to covering costs of the services they are supposed to finance (Linn et al., 1976; Bahl et al., 1976).

Francistown, Ahemdabad, Bombay, and Seoul exhibited a balanced local revenue structure, in that local taxes and self-financing service revenues contributed roughly equal shares to locally raised resources. In Francistown and Bombay this may be explained by the thriving local electricity undertakings which contributed substantially to the share of self-financed services revenues, while in Ahmedabad and Seoul it is probably a result of emphasizing benefit and user-charge financing wherever possible, even though the scope for application of such charging systems in these two cities is quite limited. Finally, in those cities, such as Tunis, where local government has least responsibility, revenue sources other than taxes or user-charges were the most important in raising local resources.

On average, approximately two-thirds of all external revenues are derived from grants and shared taxes, with the remaining third contributed by borrowing. For all cities, only 14% of all local government expenditures were financed by grants and shared taxes.[9] Thus, the degree of grant financing in cities of developing countries in general appears to be quite low and in fact lower than has typically been the case in industrialized countries (Marshall, 1969; Prud 'Homme, 1980). This contradicts Marshall's claim (1969: 41) that local governments in less-developed countries show the least degree of financial self-reliance.

The generally low shares of grant finances cast some doubt on the common belief that national governments in developing countries heavily subsidize urban, rather than rural, dwellers through their fiscal policy. If such subsidization takes place at all, it generally does not occur through transfers to urban governments. The data also suggest that one possible way out of the grave problem of urban service deficits in developing countries is to expand the degree of grant and shared-tax financing of urban governments, since this is a small source of local government finance. But public services in the cities of developing countries tend to have fewer hinterland spill-overs, since the cities do not perform as many central-place functions (education, health, entertainment, and the like) as is the case in industrialized countries.[10] Therefore, it may be quite appropriate that city governments in developing countries should receive proportionately fewer grants and transfers than in developed countries. Moreover, on equity grounds one may prefer to strengthen urban governments in raising their own resources rather than competing with the poor rural areas for scarce national government funds.

Loan financing was the smallest source of finance in Table 9.3, contributing on average only 5.7%. In this respect, cities in developing countries differ markedly from their counterparts in industrialized countries where larger capital outlays are typically financed from borrowing (U.S., HUD, 1973). This is made possible mainly because of the highly developed capital markets and the flexible regulations of higher-level governments that control local borrowing. A wide variation in the reliance on borrowing may be observed in developing countries. Apart from the two cities experiencing negative net borrowing during the periods under observation, there were four cities where no borrowing took place. The more important borrowers in this sample are the Colombian cities, the Indian cities, and Karachi. For the Colombian cities and for Bombay, this may, in part, be explained by the importance of local public utility operations, which require loan financing for their large and lumpy investments. The utilities are also good credit risks since they are often run by autonomous local agencies and financed by user charges. All of these factors tend to be looked on favorably, especially by international lenders. Also of importance is that in India, Pakistan, and Colombia the higher levels of government were relatively flexible in permitting loan financing.

For a small sample of ten cities it was also possible to determine how much the major categories of urban governmental revenues contributed to changes in per-capita expenditures over time. For those cities where urban per-capita expenditures increased in recent years, locally financed resources contributed the major share, while in those cities where urban per-capita expenditures declined, locally financed resources were the major culprit associated with this decline. To the extent that generalization is possible on the basis of this small sample, the changes in locally raised resources determine the ability of urban government to provide additional services. Where locally raised revenues fare badly, urban government expenditures suffer; where they do well, urban expenditures thrive. Central governmental transfers and borrowing play only minor roles either way.

Considering the important role played by local taxes in financing urban expenditures in developing countries, it is worth considering in somewhat greater detail the nature of the various local taxes contributing to this source of local revenue (Table 9.4). For twenty cities the percentage distribution of local tax sources is shown, with cities ordered according to the importance of locally raised revenues in financing total expenditure by each urban government. Local governments rely upon a

TABLE 9.4 Percentage Distribution of Local Tax Revenues by Source for Selected Cities

City	Year	Local Taxes as Percent of Total Local Expenditure	Property Tax	Property Transfer Tax	Income Tax	General Sales Tax	Octroi	Beer Tax	Gasoline Tax	Entertainment Tax	Industry and Commerce Tax	Motor Vehicle Tax	Gambling Tax	All Other Taxes	Total	Sources
Managua (Nicaragua)	1974	84.3	—	—	—	69.4	—	—	—	2.3	21.1	3.1	—	4.1	100.0	Lacayo et al. (1976)
Mexico City (Mexico)	1968	70.9	33.5	2.8	—	—	—	—	1.1	2.6	44.2	—	—	15.8	100.0	Fried (1972)
Valencia (Venezuela)	1968	44.8	21.4	—a	—	—	—	—	—	—	66.7	11.8	—	—	100.0	Cannon et al. (1973)
Bogota (Colombia)	1972	13.7	58.4	—	—	—	—	—a	1.8	7.0	18.2	5.1	—	9.5	100.0	World Bank Estimates
Cali (Colombia)	1974	23.2	54.0	—	—	—	—	—	—	6.1	27.8	4.3	—	7.8	100.0	World Bank Estimates
Cartagena (Colombia)	1972	23.3	61.2	—	—	—	—	—	—	4.4	12.2	2.1	5.8	14.2	100.0	Linn (1976)
La Paz (Bolivia)	1975	61.9	5.2	—	—	—	—	7.1	—	1.5	73.8	—	—	12.4	100.0	World Bank Estimates
Manila (Philippines)	1970	55.0	61.9	—	—	—	—	—	2.2	—	32.1	—	—	3.8	100.0	Bahl et al. (1976)
(Group Average)											(37.0)					
Karachi (Pakistan)	1974/5	67.6	46.0	—	—	—	49.9	—	—	—	—	3.0	—	1.0	100.0	Kee (1975)
Ahmedabad (India)	1971/2	38.6	43.0	—	—	—	52.0	—	—	—	—	2.0	—	3.0	100.0	Bahl (1975)
Bombay (India)	1971/2	37.9	55.6	—	—	—	37.7	—	—	0.3	—	3.7	—	2.7	100.0	Bougeon-Maassen (1976)
Calcutta Corporation (India)	1974/5	64.4	64.8	—	—	—	27.1b	—	—	—	—	—	—	8.2	100.0	World Bank Estimates
(Group Average)							(41.7)									
Francistown (Botswana)	1972	46.8	12.9	—	61.1	—	—	—	—	—	—	—	—	26.0	100.0	World Bank Estimates
Lusaka (Zambia)	1972	39.3	74.6	—	25.4	—	—	—	—	—	—	—	—	—	100.0	World Bank Estimates
Ndola (Zambia)	1972	n.a.	75.6	—	24.4	—	—	—	—	—	—	—	—	—	100.0	World Bank Estimates
Kitwe (Zambia)	1972	n.a.	80.0	—	20.0	—	—	—	—	—	—	—	—	—	100.0	World Bank Estimates
Kinshasa (Zaire)	1971	25.4	—	—	14.4	—	—	62.5	—	—	—	—	—	23.1	100.0	World Bank Estimates

TABLE 9.4 (continued)

City	Year	Local Taxes as Percent of Total Local Expenditure	Property Tax	Property Transfer Tax	Income Tax	General Sales Tax	Octroi	Beer Tax	Gasoline Tax	Entertainment Tax	Industry and Commerce Tax	Motor Vehicle Tax	Gambling Tax	All Other Taxes	Total	Sources
Bukaru (Zaire)	1971	67.4	–	–	3.7	–	–	87.0	–	–	–	–	–	9.3	100.0	World Bank Estimates
Mbuji-Mayi (Zaire)	1971	66.5	–	–	62.7	–	–	–	–	–	–	–	–	37.3	100.0	World Bank Estimates
Daegu (Korea)	1976	n.a.	49.5	21.2	9.1	–	–	–	–	10.4	–	5.4	–	3.5	100.0	Smith and Kim (1979)
Gwangju (Korea)	1976	n.a.	50.3	23.1	13.2	–	–	–	–	6.4	–	4.1	–	2.9	100.0	Smith and Kim (1979)
Daejeon (Korea)	1976	n.a.	51.0	20.1	9.7	–	–	–	–	10.7	–	5.5	–	3.0	100.0	Smith and Kim (1979)
Jeonju (Korea)	1976	n.a.	52.0	24.4	8.9	–	–	–	–	7.5	–	4.9	–	2.1	100.0	Smith and Kim (1979)
(Group Average)					(23.0)											
Seoul (Korea)	1971	30.3	20.6	34.8	–	–	–	–	–	16.4	–	22.2	–	6.0	100.0	Bahl and Wasylenko (1976)
Madras (India)	1975/6	54.5	68.9	5.1	–	–	–	–	–	16.0	–	–	–	10.0	100.0	World Bank Estimates
Tehran (Iran)	1974	42.8	55.3	–	–	–	–	–	–	9.1	–	10.1	–	25.6	100.0	Smith (n.d.)
Tunis (Tunisia)	1973	36.8	82.6	12.8	–	–	–	–	–	4.6	–	–	–	–	100.0	Prud'Homme (1975)
Jakarta (Indonesia)	1972/3	43.7	–a	–	–	–	–	–	–	16.9	–	50.2	26.9	6.0	100.0	Linn et al. (1975)
Lagos (Nigeria)	1962/3	50.9	100.0	–	–	–	–	–	–	–	–	–	–	–	100.0	Orewa (1966)
Kingston (Jamaica)	1972	23.9	100.0	–	–	–	–	–	–	–	–	–	–	–	100.0	Bougeon-Maassen and Linn (1975)
Rio de Janeiro (Brazil)	1967	84.4	3.9	1.0	–	89.2	–	–	–	–	–	–	–	5.9	100.0	Richardson (1973)
Overall Median		51.0														
(Overall Average)		(44.6)														

aShared tax receipts not shown under local tax revenues.
bShare of Octrol receipts collected by the Calcutta Metropolitan Development Authority.

large variety of taxes; in fact, the variety is even larger than the data in Table 9.4 suggest, since only the major taxes are shown there.

The only tax which is levied in virtually all cities is the property tax. Only two cities in the sample, Managua and Jakarta, did not levy a local property tax.[11] The median revenue contribution of the property tax to local taxes was 51 percent, and thus typically more than that of all other taxes. The only other taxes which are levied in a majority of cities are taxes on motor vehicles and on entertainment. But, neither of these two taxes contributed a substantial share of local taxes in any of the cities with the exception of the vehicle tax in Jakarta which accounted for 50% of all local taxes and approximately 22% of all local government revenues. Another tax found in a number of cities, in particular the Latin American cities and Manila, is a tax on industry and commerce. This is an important element in the local tax structure of Mexico City and Valencia, contributing approximately 30% of the financing of local government expenditure in both cities. Each of the other taxes is levied only in a few cities, but some are important in particular cases. For instance, a general sales tax was levied only in Managua, but it contributed over two-thirds to local tax revenues and almost 60% of total local expenditure financing. In some cities of India and Pakistan, a special tax called "octroi," levied on all goods entering the city boundaries, raised a significant share of local taxes. In Karachi, one-third of all local spending was financed from the octroi proceeds. Local income taxes appear to be as rare in cities of developing countries as are local general sales taxes. Two African cities (Francistown and Kitwe) retained a local income tax.[12] The category labeled "All Other Taxes" contributed a sizeable share of local taxes in some cities, such as Tehran, Francistown, Mexico, and Cartagena. Usually, the taxes falling under this heading comprise a motley collection of nuisance taxes, often costly to collect and to comply with, and that frequently show only small revenue increases from year to year.

REVENUE AUTHORITY OF URBAN GOVERNMENTS

In evaluating the ability of local governments in cities of developing countries to respond adequately to the challenge of rapidly increasing urban service needs, the degree of local authority over revenue sources is of central importance. Table 9.5 offers an overview of local government revenue authority for selected cities in developing countries.

Beginning with local taxes, the most important of these (property tax) is not freely controlled by local government. For the cities under consideration, the setting of property tax rates is always constrained by higher-level government, usually by maximum tax rate ceilings that apply to all local authorities in the country or state. Where property value assessment is a local responsibility, it is often freely handled by local agencies; the exception is Korean cities, where national and local agencies cooperate in determining property values. Other major tax bases, such as the general sales and income taxes, are rarely at the disposal of urban governments, and where they are, the authority to set rates is usually restricted by national legislation. Specific sales taxes and business taxation, where they are local options, are usually unhampered by central control, which helps explain their relative importance in cities where they are levied (especially the octroi in India and Pakistan and the industry and commerce taxes in Latin America). Motor vehicle taxes are at the disposal of urban government, but their use is generally restricted by higher-level government, with the exception of Jakarta and apparently Ahmedabad. Entertainment taxes are usually subject to restrictions by the central government, and in any case have rarely shown great potential to contribute to local revenues. Other minor taxes and fees, ironically, are the only sources of local government revenue not subject to higher-level governmental controls and restrictions. In some cities, this has led imaginative local officials to design a wide variety of these minor revenue sources, so that they add up to a noticeable proportion of total city revenues. But, they never provide a basis for healthy and reliable local revenue collections, and when local governments become too dependent on these sources of revenue, they usually face serious problems inherent in the revenue structure of the urban government.

With the exception of the Colombian and Korean cities, betterment levies and related revenue sources rarely account for important shares in revenue of cities in developing countries. But this cannot be blamed on overriding national restrictions (Table 9.5). In fact, virtually all of the cities in this sample were

TABLE 9.5 Revenue Authority of Urban Governments by Revenue Source and City

Revenue Source	Ahmedabad	Bombay	Bogota	Cali	Cartagena	Jakarta	Manila	Seoul	Daegu Gwangju Daejeon Jeonju	Nairobi	Lusaka Ndola Kitwe	Kingston
Taxes												
Property Tax												
Assessment	F	F	F	-	-	-	F	R	R	F	R	-[d]
Tax Rate	R	R	R	R	R	-	R	R	R	R	R	-
General Sales Tax	-	-	-	-	-	-	-	-	-	-	-	-
Specific Sales Tax	F	F	-	-	-	-	F	-	-	-	-	-
Income Tax	-	-	-	-	-	-	-	R[a]	R[a]	-	R	-
Business Taxes	-	-	F	F	F	F	R	R[b]	R[b]	F	-	-
Vehicle Taxes	F	R	R	R	R	F	-	R	R	-	-	-
Entertainment Taxes	F	R	R	R	R	F	-	(R)[c]	(R)[c]	-	-	-
Minor Taxes and Fees	F	F	F	F	F	F	R	F	R	F	R	R
Betterment Levies	F	F	F	F	F	F	F	F	F	-	-	F
User Charges												
Water and Sewerage	F	F	R	R	R	F	R	R	R	R	R	-
Electricity	-	F	R	R	-	-	R	-	-	-	-	-
Telephones	-	-	R	R	R	-	-	-	-	-	-	-
Housing	-	F	F	F	-	F	F	F	R	R	F	-
Public Transport	-	F	R	-	-	F	-	R	-	-	-	-
Shared Taxes	N	-	N	N	N	N	N	N	N	-	-	-
Grants	N	N	N	N	N	N	N	N	N	N	N	N
Borrowing												
Private Capital	R	R	R	R	R	R	R	R	R	R	R	R
Public Capital	R	R	R	R	R	R	R	R	R	R	R	R
Sources	Bahl (1975)	Bougeon-Maassen (1976)	Linn (1980)	Linn (1980)	Linn (1980)	Linn et al. (1976)	Bahl et al. (1976)	Bahl and Wasylenko (1976)	Smith and Kim (1979)	World Bank	World Bank	Bougeon-Maassen and Linn (1975)

[a] Introduced in 1973.
[b] Introduced in 1976.
[c] Abolished in 1976.
[d] Local property taxation was abolished in 1974.

LEGEND:
F: Freely administered at local level.
R: Subject to restrictions by higher-level government.
N: Administered by national (or other higher-level) government.
-: Revenue source not utilized at local level.

entitled to pursue freely this form of financing. The reasons for the failure to use this revenue source must, therefore, be found in technical problems associated with its application, in the reluctance of local authorities to apply these levies, or in lack of familiarity with the principles and practices of betterment levies.

User charges are often controlled by central authorities. In some countries, for instance in Colombia, these controls include nationwide guidelines and review of changes in user charges. In Colombia, national-level intervention has supported pricing according to average long-run costs of public utility services (Linn, 1980a). In other countries and cities, the central control over local user charges has had negative effects on local revenue efforts. Some of the Korean cities, for instance, were unable to raise their user fees to levels thought appropriate by local authorities due to higher-level governmental interventions (Kim and Smith, 1979). The only major self-financing service where local governments are generally free to set charges is the provision of public housing. But to the extent that national government is involved in the planning and design of local housing projects and through capital contributions and approval requirements for borrowed funds, it may also influence the cost recovery policies. In any case, in virtually all cities where major local housing programs are carried out severe financial problems typically have plagued the local authorities (e.g., in Zambian cities) due to subsidized housing operations.

External revenue sources, not surprisingly, are even less amenable to local control than are sources of locally raised revenues. Local governments never appear to have a say in the determination of shared taxes and grants. In particular, it is of importance that the practice of "piggy-backing" local taxes on national taxes[13] does not appear to exist anywhere in developing countries.

Local borrowing appears always to be subject to considerable restrictions by higher-level authorities. In some countries, such as Colombia, the planning ministry has to review and approve all applications for borrowing exceeding a small ceiling. Below this ceiling, local authorities may borrow freely from commercial banks, mainly for cash-flow management (Bird, 1975). In other countries, e.g. Kenya, local public agencies have to receive prior authorization for all borrowing. In yet other countries (e.g., Algeria, Sri Lanka, Malaysia, Pakistan, Sudan, and Egypt), local authorities may only borrow from special loan funds set up and controlled by higher-level governments (Marshall, 1969).[14]

These national restrictions on borrowing by local governments are not surprising in countries where capital markets are poorly developed and where the national government has to be concerned about the nationwide allocation of public and private savings, and therefore tries to channel public investments, including those by local authorities, to conform with the objectives and directions of the national plan. Industrialized countries do not face similarly severe constraints on their capital resources and thus tend to control local borrowing much less. But even there, the unconstrained access of local authorities in the United States to private captital markets is exceptional.[15]

The combination of poorly developed capital markets, severe savings constraints, and higher-level government restrictions and controls over local government borrowing explains why this source of finance for urban services is the least important of all sources of local government revenue and much less important in LDCs than in industrialized nations.[16] These constraints on borrowing are particularly troublesome where major lumpy investment projects need to be financed by local authorities, particularly in the area of public utilities, but also for slum-improvement, public housing, and sites-and-service projects, or major school construction projects. Where borrowing is impossible, the investments have to be financed by grants or have to be made by higher-level government agencies, or, more frequently yet, they cannot be made at all.

In summary, urban governments in developing countries generally do not have access to the more buoyant and flexible taxes, such as income taxation, general sales taxes, or tariffs on international trade. Instead, they are usually entitled to levy a tax on real estate and some combination of taxes on the sale of particular commodities or on business activity within the urban area. Minor taxes, including those on various forms of entertainment or gambling, are often the only recourse for independent tax efforts at the local level. But even where local government enjoys substantial freedom of action, it frequently has not used existing revenue authority to the extent feasible. The most important of the underused revenue instruments are motor vehicle taxation, betterment levies, and user charges. Central governmental control and intervention are neither necessarily a help to local authorities in raising revenues nor necessarily a barrier. What matters is how the national (state) authorities see their roles in relation to local authorities: Where higher-level government is supportive, as, for instance, in Colombian cities and in Seoul, local government can improve its reliance on its own resources even under a fairly restricted framework. On the other hand, where higher-level authorities view their role as competitive and as one of controlling and limiting local government efforts in raising additional revenues, higher-level authorities are likely to inhibit severely the revenue raising capacity of local government.

ISSUES AND PROBLEMS OF URBAN FINANCES IN DEVELOPING COUNTRIES

The preceding section pieced together a picture of the prevailing patterns of expenditure responsibility and revenue authority of urban governments in developing countries on the basis of financial data and institutional information collected for a number of cities. The present section goes beyond the available governmental statistics in considering some of the prime issues and problems of urban finances in developing countries. To a considerable extent, these issues and problems are directly related to the broader issues of economic development and urbanization that are, however, beyond the scope of this paper (World Bank, 1979). Not surprisingly, the issues and problems confronting urban governments in developing countries are different than those confronting cities in the industrialized nations, despite the fact that the rhetoric surrounding urban finance problems is often similar in both settings. The following discussion briefly characterizes these differences.

THE EFFICIENCY OF URBAN GOVERNMENT

Urban governments in developing and developed countries have been criticized for an inefficient allocation of resources. Typically, in the developed countries the claim is that the urban authorities are too heavily involved in the lives of their citizens, while in the developing countries urban governments are challenged for not providing enough support. The following exposition may help to put these concerns into a perspective, directing attention to the more important areas where fiscal policy can and should intervene in adjusting the shortcomings of prevalent urban finance systems.

Consider first an ideal urban economy where the production of privately and publicly provided goods and services takes place efficiently and where the distribution between the consumption of these two types of commodities is efficient. In diagrammatic terms, this means that the urban economy is located on point A of the production possibility frontier (transformation curve) P in Figure 9.1, where the community indifference curve I is tangential to P.

But now assume that instead of efficiently representing community preferences, the decision makers have selected consumption points B or C. In the former case, too little publicly produced goods are provided—in the latter, too much. The allocation of resources is thus inefficient because the new consumption and production points (B and C) lie on a community indifference curve II below the highest feasible curve I. Since B and C are still on the production possibility frontier P, the production of private and public goods remains efficient. The misallocation of resources enters on the consumption side through the mistaken choice of consumption along P. To get to the efficient consumption point A, an increase in publicly produced goods and a reduction in privately produced goods would have to take place if the consumption point is B, and the converse, if the consumption point is C. The concern over excess-provision of publicly produced goods and services in some of the cities of developed countries, may, in large part, involve the judgment that the city is at a point like C and that

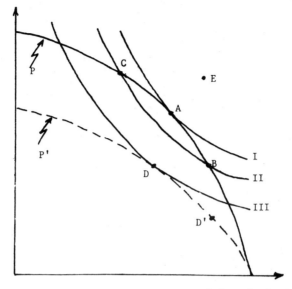

Figure 9.1 Production and Consumption of Privately and Publicly Provided Goods and Services

for efficiency a reduction in the relative size of the public sector, and a concommitant increase in the private sector, are desirable.

In developing countries, one frequently hears the argument that urban public services are underprovided. This might imply that the urban economy is situated at a point such as B, and that, therefore, a shift from privately to publicly produced goods and services is desirable by moving along P from B to A. But, it is possible, and probably more appropriate, to give a different interpretation to the concern with urban finances in developing countries. Assuming that the public sector is inefficient in its production of goods and services, it is possible to draw a production possibility frontier P', which lies to the left and below of P.[17] Assuming, moreover, that the policy makers choose the efficient consumption point on P', this must be D where community indifference curve III is tangential to P'. In this case, the problem is not an underprovision of publicly produced services due to a mistaken consumption decision, but the problem lies in the inefficient production of public services. The solution must therefore be found in the inefficiencies in public production. This would permit the urban economy to move from D to A by shifting the production possibility frontier from P' to P. As a result, an increased amount of publicly produced goods and services would be provided while, at the same time, an expansion in private sector production would also occur. In these circumstances, the expansion of the public sector does not result in a concommitant reduction in the private sector.

It is of course, possible that cities of developing countries are not at the efficient consumption point D on P', but they may be at point D' where publicly produced good are underprovided along the inefficient production possibility frontier. An urban public finance reform should, therefore, aim simultaneously at eliminating the inefficiencies in the public sector and at correcting the bias in consumption in favor of privately produced commodities. This would require a stronger expansion of the public sector than would have been the case had the economy been located at D, rather than D'. But, the crucial aspect is that a shift alone in production from the private to the public sector is not sufficient to reap the entire possible increase in welfare. An effort must also be made to improve efficiency of public sector production. Better management and training, use of appropriate low-cost technologies, and more effective coordination of public agencies are among the methods that can be used to that purpose.

THE URBAN FISCAL GAP

Urban governments in developing and developed countries alike complain about a lack of resources to provide sufficient urban services. The gap between perceived service needs and financial resources, also referred to as the fiscal gap, is subject to various interpretations. One way of interpreting its meaning can be exemplified by Figure 9.1. The desired level of public expenditures may involve a combination of public and private production reflected by point E, which lies beyond the efficient production possibility frontier. By definition, this point is beyond the resources available to the urban economy even where they are allocated in the most efficient manner. There is, consequently, a gap between perceived public expenditure needs and available resources. One common response among urban governments in developing countries has been to attempt to provide public services at standards and levels commensurate with the requirements of E. But this has generally resulted in many citizens having to do without public services altogether and, therefore, led to an inefficient allocation of public sector resources. The urban economy thus ended up on a lower production possibility curve than would have been the case with standards of service provision commensurate with public goods production at a point such as B. This analysis is important because it points to the need to assess realistically the resource constraints under which an urban economy operates in determining public service needs and standards.

But the notion of a fiscal gap can be given another interpretation. Urban populations in most of the developing countries have expanded rapidly in recent years and are likely to continue to do so in the future (World Bank, 1979). This growth has led to rapid increases in expenditure requirements because of the need to provide each new urban dweller with some minimum level of public services, and also possibly because of increases in costs associated with urbanization. Furthermore, rising incomes in urban areas also increase the demands for public services (Linn, 1980b). But at a time when urban expenditure requirements increase rapidly, the revenues of local authorities most directly affected by these increases in expenditure needs generally do not increase commensurately. The reason for this is that urban governments are often restricted in their revenue authority to relatively inelastic revenue sources. The fiscal gap of urban governments interpreted thusly is a result of a mismatch between expenditure responsibilities and revenue authority of urban governments.

Fiscal reform can do little to "solve" the first kind of fiscal gap except to ensure that the public sector functions efficiently and is not debilitated by unrealistic expectations. Reform is possible for the second type and should provide urban authorities with sources of revenue that have the potential to grow with urban population and incomes—unless the higher-level government is willing and able to assume responsibility for the growing urban service demand.

FISCAL CENTRALIZATION VERSUS DECENTRALIZATION

In attempting to deal with the fiscal gaps experienced by urban governments, national authorities thus inevitably have confronted the issue of whether to increase the extent of fiscal centralization. In fact, one common response of higher-level government has been to assume the responsibility for certain urban services, thus reducing some of the pressures on the local authorities, but at the same time also reducing the local government's scope to respond to urban policy issues and problems. Quite frequently, national governments have assumed direct responsibility for carrying out capital investments in such areas as public utilities, roads, education, and health, while leaving local authorities with the responsibility to operate and maintain the physical plants and facilities once constructed. This "turnkey" approach has significant disadvantages in that it tends to burden local authorities with facilities often beyond the local financial and technical ability to operate and maintain and may well not reflect local preferences. It also virtually ensures that capital costs are financed from the national budget, rather than from user charges.

Another common response has been to provide local authorities with a patchwork of temporary financial assistance to bail them out of the most urgent financial and political crises. Contrary to the experience in developed countries, national governments in the developing countries have been reluctant to provide local authorities with substantially increased revenue authority or a permanent and buoyant revenue sharing system. The reason for this reluctance generally has been founded in the fear that local authorities are unfit to use increased resources effectively or would use them for purposes contrary to the interests of the national government. This lack of confidence of higher-level governments in their lower-level counterparts is pervasive in many developing countries and has foiled most attempts to increase fiscal decentralization, as well as most attempts to come to grips with the fiscal gaps of urban governments. By implication, many programs aimed at increased decentralization of economic developing have also failed, because they did not strengthen sufficiently the local authorities in the small- and medium-sized towns, which were to attract an increasing share of economic growth away from the largest cities.

Whatever the preference of national policy makers in terms of more centralized or decentralized administrative and economic development, effective urban service provision requires effective and mutually supportive cooperation between all levels of government. The sense of competition, suspicion, and lack of confidence frequently prevailing between national and local governments in developing countries must be reduced wherever possible. Experiments with the municipal development agencies found in some Central and South American countries (Gall, 1976) provide useful examples of how to develop national institutions and programs in support of local authorities.

Urban governments, particularly those in the larger cities, should be encouraged and supported to develop their own financial resources to execute their functions, whether large or small. An increasing reliance on charges and taxes imposed on the beneficiaries of urban services and related more or less directly to the costs of service provision is likely to improve the efficiency, equity, and fiscal viability of urban government. Increased financial self-reliance of city governments does not rule out selective transfer schemes, particularly where these are designed to assist those local authorities suffering from low revenue bases (especially in smaller towns and rural areas) or facing special expenditure responsibilities as is sometimes the case with capital cities.

FISCAL FRAGMENTATION

Another common response to the urban fiscal problems in developing countries has been the establishment of autonomous agencies, particularly for the purpose of providing urban infrastructure services. These agencies are generally endowed with flexible and buoyant revenue sources and more independence in managing their own affairs than the local government from which they were separated. The proliferation of such autonomous agencies with often overlapping functions makes difficult the systematic and rational provision of services, although there may be gains in increased professionalism and managerial effectiveness within each of the autonomous agencies. In Bogota, for example, some 15 independent local public agencies as well as various national government ministries, are involved in transportation, housing, education, and health (World Bank, 1979). Unnecessary duplication in service, inefficient sequencing of investment programs, and lack of concern for the broader implications of each agency's actions reduce the overall efficiency of the urban government. Moreover, those service functions not favored by the creation of strong, single-purpose agencies are usually underprovided by the weak general purpose authority that retains residual service responsibility.

In contrast to this functional fragmentation, geographic fragmentation of urban government within metropolitan areas has not been a problem in most developing countries. But, this problem, which has been serious in the United States, will also trouble more and more of the larger cities in the developing countries as rapid urban population growth spills over into adjacent jurisdictions. This problem has already occurred in Mexico City and awaits a resolution.

Functional and geographic deconcentration of public service responsibility within metropolitan areas may have advantages where it results in effective cost savings or gains in a more accurate reflection of differential neighborhood preferences in service provision. It is not clear, however, that the prevailing patterns of fragmented public responsibility are the rational response to achieve either of these goals. A reduction in functional and geographic fragmentation, or at least efforts to prevent further fragmentation, are likely to be desirable in many cities of developing countries.

THE SCOPE AND PROSPECTS FOR REFORM

Fiscal reform proposals have been made in every major city and in every country to alleviate the serious problems facing urban governments. While the nature of these reform proposals has varied from place to place in line with local conditions and with each study team responsible for the proposals, only few major reforms were actually made in response to reform proposals (Walsh, 1969; Marshall, 1969; Robson and Regan, 1972; United Nations, 1975).[18] Those countries where major reforms have taken place in the last twenty years are mainly among the industrialized countries, e.g., Germany (consolidation of communes and reform of revenue sharing arrangements); Sweden (consolidation of communes); and the United States (reform of revenue sharing arrangements). Among the developing countries, Yugoslavia is probably the main exception to the rule that changes in urban finance arrangements are very slow to proceed, and that it may take decades for fundamental changes to take place, if they occur at all (Walsh, 1969; Pusic and Walsh, 1968).

Much more typical for the developing countries are minor and slow adjustments in urban finance practices, such as the creation of special districts for capital cities with special expenditure responsibilities and revenue authority (Bogota and Seoul); enlargement of metropolitan jurisdictions by annexation of adjacent municipalities (Bogota); gradual development of new revenue sources (betterment levies in Colombian cities, land readjustment schemes in Korea, and vehicle taxation in Jakarta); gradual reform of existing revenue sources (property taxation in Jakarta); minor reassignments of expenditure functions (Kenya and Zambia); and similar gradual and ad hoc responses to urban fiscal pressures. Major reform proposals were typically shelved or taken up only in very minor respects, e.g., local government reform proposals in India and property tax reform in Jakarta. Where major adjustments in the fiscal structure of urban governments have occurred, it was either where higher government took over important revenue sources previously allocated to local authorities (Kenya and Iran); where sweeping political changes resulted in major shifts of national policy priorities (Nigeria, Tanzania, and Uganda); or where the fiscal problems became so unmanageable that some drastic reform was unavoidable (removal of the most important expenditure responsibilities from rural and small town councils in Kenya).

The lesson from the history of urban fiscal reform proposals and implementation is that major proposals rarely have a chance for adoption and implementation, and that gradual and stepwise adjustments of the existing structure toward a more desirable state is all that can be hoped for, and may in fact have more chance of eventual implementation. There exist two major reasons for the inertia typically found in the face of, the need for, urban public finance reform: First, policy makers and citizens share a natural antipathy to the uncertain effects of untested, large-scale changes in the economic environment; and, second, most major reforms are associated with substantial windfall losses to relatively few among the urban population—mostly among the elites—while windfall gains are likely to be spread over a much larger number of people—mostly among the less well-off.

NOTES

1. One of the four theme papers presented at the third session of the U.N. Commission on Human Settlements in Mexico City in 1980 regretfully notes the lack of attention given to local service financing (United Nations, 1980).

2. Throughout this study, local government is defined as the lowest level of the standard three-tier governmental structure which distinguishes between national, state (province), and local authorities. Subnational government is defined to cover the local and state (province) levels of government. In many smaller countries, no intermediate-level governments exists.

3. While Kee's figures provide a general impression of the relative importance of various levels of local government, they have to be treated with caution due to problems of data and definition. Few developing countries have universal reporting for all subnational authorities and therefore estimates of local and state spending may vary between different sources. Also, the definition of the government sector varies between countries and analysts. Of particular importance is whether or not autonomous public agencies and enterprises are defined to be part of the government sector.

4. Again, caution must be exercised in the analysis of these data. While the underlying expenditure figures for the local authorities are derived directly from the local budgets, and are therefore quite reliable, the spending figures for higher level government are, with the exception of Cali, Mexico City, and Karachi, based on the assumption that the national (state) government spends as much on a per capita basis in the city as it does on average throughout the country (state). For Cali, Mexico City, and Karachi direct estimates of higher level government spending in each city were available. However, even these are subject to limitations, since it is always difficult to determine what shares of central government expenditures should be allocated to particular urban areas.

5. This is generally not done in comparative analyses of local government expenditure, due to limitations on the available data. In the present context, case

studies were relied upon where an effort had been made to capture the expenditures of all major local government agencies.

6. One of the serious problems in evaluating the fiscal performance of urban governments in developing countries is the typically very poor separation of current from capital expenditures, making the evaluation of spending patterns very difficult, both for purposes of policy analysis in a given city, and for cross-city and cross-country comparisons.

7. For lack of space these figures are not reproduced here.

8. Locally raised revenues can finance more than 100% of total local government spending where local authorities run substantial surpluses. The resulting accumulation of resources (either in the form of liquid funds or of financial assets) is then shown as negative net borrowing.

9. This is below the estimate of Walsh (1969), who estimated that on average 20% to 25% of local revenues are derived from grants. Similarly, on the basis of Smith's data (Smith, 1974) an average of 19.6% of local government revenues in cities of developing countries are contributed by grants and shared taxes. The lower average shown in the present study may in part be explained by the different sample of cities, but it probably also results from a more comprehensive definition of local government, which includes the autonomous and semi-autonomous local service enterprises that are generally not included in other analyses of local government finances.

10. The possible reason for this hypothesis is that the automobile is not nearly as common in the rural areas in developing countries as it is in industrialized countries, thus resulting in a much lower mobility of the rural population (short of outright migration to the city).

11. Jakarta draws on the property tax as a shared tax that is now largely under the control of the central government, and therefore, does not show up in Table 9.4. The local government in Kingston lost the authority to collect the property tax in 1974, as did all other local governments in Jamaica.

12. Nairobi also had an important local income tax, until it was abolished by the central government in 1973. Korean local governments were given the power in 1973 to raise a combination of local head and income taxes (Smith and Kim, 1979).

13. This is a form of shared taxation, where national (state) and local taxes on the identical base are collected by the national (state) government, in order to save on administration and collection costs, but where local governments have the freedom to determine the rate at which the local tax is levied in their respective jurisdictions.

14. In fact, the line between higher-level grants and loans frequently becomes blurred, since the public loans are often provided at highly subsized rates and sometimes jointly with grants in financing local government capital spending (e.g., in India, Pakistan, and Jamaica).

15. In Austria, for instance, there exist nationally set ceilings on local borrowing expressed as a percentage of revenues raised locally by the local governments (Bauer, 1971).

16. One particular problem with extensive central control over local borrowing is the often excessive paper work and delays which are encountered when having to apply for permission to borrow even relatively minor amounts. This in itself has helped to discourage borrowing by local governments; see, e.g., Bird (1975) for Cali.

17. The private sector may also be operating inefficiently. But these inefficiencies are largely beyond the control of the local authorities and are therefore not explicitly considered here.

18. Many of the case studies of urban finances cited in this paper provide additional support for this contention.

REFERENCES

APEL, H. (1977) "Wie lang soll Bonn die Zeche zahlen?" Die Zeit, 15 July.
BAHL, R. W. (1975) "Urban public finances in developing countries: a case study of metropolitan Ahmedabad." Urban and Regional Report 77-4. Washington, DC: World Bank.
———, P. BRIGG, and R. S. SMITH (1976) "Urban public finances in developing countries: a case study of metropolitan Manila." Urban and Regional Report 77-8. Washington, DC: World Bank.
BAHL, R. W. and M. WASYLENKO (1976) "Urban public finances in developing countries: a case study of Seoul, Korea." Urban and Regional Report 77-3. Washington, DC: World Bank.
BAUER, H. (1971) "Finanzen," in E. Matzner (ed.) Wirtschaft und Finanzen österreichischer Städte. Vienna: Institut für Stadtforschung.
BEIER, G., A. CHURCHILL, M. COHEN, and B. RENAUD (1976) "The task ahead for the cities of developing countries." World Development 4: 363-409.
BIRD, R. M. (1975) "Intergovernmental fiscal relations in a developing country: the case of Cali, Colombia." World Bank, Washington, DC (mimeo).
BOUGEON-MAASSEN, F. (1976) "Urban public finances in developing countries: a case study of metropolitan Bombay." World Bank, Washington, DC (mimeo).
——— and J. F. LINN (1975) "Urban public finances in developing countries: a case study of metropolitan Kingston, Jamaica." Urban and Regional Report 77-7. Washington, DC: World Bank.
CANNON, M. W., R. S. FOSLER, and R. WITHERSPOON (1973) Urban Government for Valencia, Venezuela. New York: Praeger.
EHRLICHER, W. and R. HAGEMANN (1976) "Die öffentlichen Finanzen der Bundesrepublik im Jahre 1974." Finanzarchiv (N.F.) 35: 322-346.
FRIED, R. C. (1972) "Mexico City," in W. A. Robson and D. E. Regan (eds.) Great Cities of the Third World. London: Allen & Unwin.
GALL, P. M. (1976) Municipal Development Programs in Latin America. New York: Praeger.
GOLDFINGER, C. (1975) "Municipal finances in Gujranwala." World Bank, Washington, DC (mimeo).
KEE, WOO SIK (1976) "Fiscal decentralization and economic development." World Bank, Washington, DC (mimeo).
——— (1975) "Government finances and resource mobilization in Pakistan." World Bank Studies in Domestic Finance 8. Washington, DC: World Bank.
LACAYO DE ARGUELLO, R. D., G. L. WONG, and J. V. ARBOLEDA (1976) "Finanzas publicas locales de Managua: estructura de ingresos." Managua: Vice Ministerio de Planificacion Urbana (mimeo).
LINN, J. F. (1980a) "The distributive effects of local government finances in Colombia: a review of the evidence," in R. A. Berry and R. Soligo (eds.) Economic Policy and Income Distribution in Colombia. Boulder, CO: Westview.
——— (1980b) "The costs of urbanization." Urban and Regional Report 79-16. Washington, DC: World Bank.
——— (1979) "Policies for efficient and equitable growth of cities in developing countries." Washington, DC, World Bank Staff Working Paper 342.
——— (1975) "Urban public finances in developing countries: a case study of Cartagena, Colombia." Urban and Regional Report 77-1. Washington, DC: World Bank.

———, R. S. SMITH, and H. WIGNJOWIJOTO (1976) "Urban public finances in developing countries: a case study of Jakarta, Indonesia." World Bank, Washington, DC (mimeo).
MARSHALL, A. H. (1969) Local Government Finance. The Hague: International Union of Local Authorities.
OREWA, G. O. (1966) Local Government Finance in Nigeria. Ibadan: Oxford University Press.
PRUD 'HOMME, R. (1980) "Fiscal issues of metropolitan areas." Institut d' Urbanisme de Paris.
——— (1975) "Urban public finances in developing countries: a case study of metropolitan Tunis." Urban and Regional Report 77-2. Washington, DC: World Bank.
PUSIC, E. and A. H. WALSH (1968) Urban Government for Zagreb, Yugoslavia. New York: Praeger.
RENAUD, B. (1979) "National urbanization policies in developing countries." Washington, DC, World Bank Staff Working Paper 347.
RICHARDSON, I. L. (1973) Urban Government for Rio de Janeiro. New York: Praeger.
ROBSON, W. A. and D. E. REGAN (1972) Great Cities of the World: Their Government, Politics and Planning. London: Allen & Unwin.
SMITH, R. S. (1974) "Financing cities in developing countries." IMF Staff Paper 21: 329-388.
——— and C.-I. KIM (1979) "Local finances in non-metropolitan cities of Korea." Urban and Regional Report 79-1. Washington, DC: World Bank.
SMITH, T. R. (n.d.) "Tehran's fiscal structure." (Mimeo).
United Nations (1980) "Human settlements finance and management." Theme Paper for the third session of the United Nations Commission on Human Settlements, Mexico City, 6-14 May.
——— (1975) Local Government Reform: Analysis of Experience in Selected Countries. Department of Economic and Social Affairs. New York: United Nations.
U.S. Department of Housing and Urban Development (1973) A Study of the Financial Practices of Governments in Metropolitan Areas. Washington, DC: Government Printing Office.
WALSH, A. H. (1969) The Urban Challenge to Government. New York: Praeger.
World Bank (1979) World Development Report, 1979. Washington, DC: World Bank.

The Contributors

ROY BAHL is Director of the Metropolitan Studies Program and Professor of Economics and Public Administration in The Maxwell School at Syracuse University. He has written widely on matters of state and local government finance and urban regional economics, and he is a frequent adviser to governments and the private sector in this country and abroad.

ROBERT J. CLARK is currently a policy analyst with the Southern Growth Policies Board. He attended the University of Oklahoma, where he obtained his Bachelor's degree in citizenship and public affairs. He received his Master's degree in political economy from the University of Texas at Dallas.

JAMES W. FOSSETT is a doctoral candidate in political science at the University of Michigan and a staff member of the Princeton Urban and Regional Research Center. He has been a staff member of The Brookings Institution and an official in Georgia's state government. He has published several papers on urban politics and policy.

SHAWNA GROSSKOPF is Assistant Professor of Economics at Southern Illinois University. Her interest in public employment was initiated at Syracuse University while working on *Public Employment and State and Local Government Finance.* Other areas of interest include the theory of clubs and public utility pricing.

JOHANNES F. LINN received his Ph.D. in economics from Cornell University. He currently serves as a senior economist in the Development Economics Department of the World Bank. His primary field of research has been the comparative analysis of urban government finances in developing countries, with particular emphasis on local taxes and service charges. He has also published various papers on urban and regional development issues in developing countries, and has worked on problems in the economic analysis of investment projects.

JOERGEN R. LOTZ is Deputy Director of Finance for the city of Copenhagen. He formerly served as an economist in the Danish Ministry of Taxation, in the Fiscal Affairs Department of the International Monetary Fund, at OECD, and as a lecturer at the University of Copenhagen. He has written many papers in the areas of taxation, fiscal policy, and local government finance.

DEBORAH MATZ is an economist with the U.S. Congress, Joint Economic Committee. She is responsible for preparing committee reports and hearings, and for conducting and managing studies on issues concerning state and local government finance, intergovernmental relations, economic and industrial development, housing, and related areas. In addition to advising members of Congress on economic policy affecting these areas, she directs the initiatives of the Joint Economic Committee's Subcommittee on Fiscal and Intergovernmental Policy.

RICHARD P. NATHAN is Professor of Public and International Affairs at The Woodrow Wilson School of Princeton University and directs the Princeton Urban and Regional Research Center. Until 1979 he was a senior fellow at The Brookings Institution and an official of the U.S. Office of Management and Budget, and the Department of Health, Education and Welfare.

ROBERT REISCHAUER is the Deputy Director of the Congressional Budget Office. He is an economist and has written in the areas of state and local fiscal problems, the economics of education, revenue sharing, welfare policies, public finance, and the federal budget. He formerly served as a staff member of the economic studies division of The Brookings Institution.

MICHAEL WASYLENKO is Assistant Professor of Economics at Pennsylvania State University and, on academic leave during 1980-1981, a Visiting Scholar in the Office of Policy Development and Research at U.S. Department of Housing and Urban Development. He has published articles on firm location choice and on state and local government expenditure and taxation. He has also served as a member of Governor Thornburgh's 1980 Commission on Tax Reform in Pennsylvania.

BERNARD L. WEINSTEIN is Professor of Economics and Political Economy at the University of Texas at Dallas, where he also serves as Associate Director of the Center for Policy Studies. He served as scholar-in-residence with the Southern Growth Policies Board in Washington, D.C., from September 1978 to August 1979, and as the board's Associate Director for Federal Affairs from 1979 to 1980. A consultant to many public and private organizations, he is also directing the economics task force of the 1980 Commission on the Future of the South.